Cotton pests and their control in the Near East

FAO
PLANT
PRODUCTION
PROTECTION
PAPER

41

Report of an FAO Expert Consultation
Izmir, Turkey
5-9 September 1994

Food
and
Agriculture
Organization
of
the
United
Nations

Rome, 1997

M-15
ISBN 92-5-103799-X

Foreword

Owing to its comlex nature as a fibre crop, an oil crop and a feed crop, as well as to increased market demand and its high value, cotton is an economically important agricultural crop for policy-makers, consumers and farmers. As a result, it receives government and industry attention. In most Near Eastern countries, cotton cultivation, protection and marketing are the sole responsibility of the government.

In the Near East, cotton is grown as an economic crop in 11 countries –Afghanistan, Egypt, the Islamic Republic of Iran, Iraq, Morocco, Pakistan, Somalia, the Sudan, the Syrian Arab Republic, Turkey and Yemen. The area at present given over to cotton production in these countries is about 4.5 million hectares or about 12 percent of the total world cotton area. This area produces about 9 million tonnes, representing over 16 percent of total world seed cotton production. Cotton production in Egypt, Pakistan, the Sudan and Turkey represents about 85 percent of the total cotton production of the Near East region.

Among the limiting factors to cotton production in the Near East is seasonal pest infestation. Cotton is subject to attack from more than 20 economically important arthropod pests, 12 disease agents and 22 noxious weeds. This means that cotton protection measures have to be taken. In most cases, pesticide use has been the sole method used for cotton protection in Near Eastern countries, but this has resulted in complications such as pesticide residues in cotton, pest resistance to pesticides, the emergence of previously minor pests as major pests, disturbances in the natural balance, pesticide poisoning and high cotton production costs.

These complications are affecting the sustainability of cotton production in the region and have become a major concern to policy-makers as well as to the various sectors involved in cotton production, industry and marketing.

To assist its member countries in defining alternative efficient, safe and cost-effective control methods to combat the cotton pest complex, FAO invited experts in cotton protection from the major cotton producing countries in the region, as well as experts in the implementation of successful cotton integrated pest management programmes from countries outside the region, to meet and discuss cotton pest problems in the Near East and the impact of the applied control measures, and to propose an integrated pest management strategy that addresses the cotton pest problem and contributes to the sustainability of the crop in the Near East region.

This publication outlines the status of cotton pest problems and the applied control measures practised in the Near East region and furnishes the outcome of the meeting. FAO encourages its member countries to implement the proposed strategy and then to evaluate it and improve it as necessary.

M.M. Taher
Senior Plant Protection Officer
FAO Regional Office for the Near East

Contents

Editor's note

At the conclusion of the FAO Expert Consultation on Cotton Pests and their Control in the Near East Region a short report was prepared and unanimously adopted by the participants. The report was subsequently distributed by the FAO Regional Office for the Near East. The proceedings of the consultation presented in the following represent an expansion of the earlier report, in that the individual country reports and, where available, the invited papers are now given in full. However, to avoid unnecessary repetition the earlier account of the round-table discussions has been condensed somewhat, although it is hoped that the substance of what was said has been retained.

The country reports are an important part of these proceedings. They all follow the same general format but differ in the emphasis and amount of detail given to various aspects of cotton pest management. These differences reflect significant differences among the countries in their experiences of pest problems on cotton and the manners in which they have sought to overcome these problems. Each country report has therefore been edited so that it may be read as a self-contained document without the need to refer to the others.

This approach has meant a certain amount of repetition especially in matters such as the use of the full scientific name of each pest at its first mention in each report. As far as possible these names have been checked with the authorities acknowledged below, although any errors remain the responsibility of the editor. In a few cases it was found that the names of pests given in the country reports have been superseded and so the currently preferred names have been substituted. English names of arthropods and pathogens follow those given in Matthews and Tunstall (1994) and Hillocks (1992) respectively. There appear to be few commonly accepted English names for the weeds occurring in the region so none is used.

There was considerable variation in the way pesticide-related information was presented in the country reports and it has not always been possible to make full use of the information in these proceedings. This serves to emphasize the need to use the International Organization for Standardization (ISO) common name (rather than local trade names) and the International Group of National Associations of Agrochemical Manufacturers (GIFAP) code for formulation type (GIFAP, 1989), in publications on pesticides, as well as giving details of the concentration of active ingredient in the formulation. Application rates are best expressed as grams of active ingredient per hectare. If this is done ambiguities should be eliminated and comparisons with data from other sources facilitated, even if in practice other units, for example acres of feddans, were sometimes used during the application. In the following, to enable comparisons to be made among countries, data have as far as possible been converted to metric units. Pesticide nomenclature follows that given in Tomlin (1994).

D.J. de B. Lyon
October 1995

References

GIFAP. 1989. *Catalogue of pesticide formulation types and international coding system.* Technical monograph No. 2, February 1989. Brussels, Belgium, GIFAP.

Hillocks, R.J. ed. 1992. *Cotton diseases.* Wallingford, UK, CAB International.

Matthews, G.A. & Tunstall, J.P. eds. *Insect pests of cotton.* Wallingford, UK, CAB International.

Tomlin, C. ed. 1994. *The pesticide manual,* 10th edition. Farnham, UK, British Crop Protection Council and Cambridge, UK, the Royal Society of Chemistry.

Acknowledgements

The Editor wishes to thank:
Mr L. McVeigh, Overseas Development
Administration (ODA), Natural
Resources Institute (NRI),
for technical advice;
Mr J. Terry, Head of the Tropical Weeds
Unit, Long Ashton Research Station,
Bristol, United Kingdom, for advice on
weed nomenclature;
Mr J.M. Waller, Commonwealth
Agriculture Bureau International
(CABI), Mycological Institute,
for advice on pathogen nomenclature;
Dr J. Bridge, CABI, Institute of
Parasitology, for advice on nematode
nomenclature;
Mr D. Girling, CABI, Institute of
Biological Control, for checking insect
and mite nomenclature on CABI's
Arthropod Name Index on CD-rom;
and Mr Michael Goulding for word-
processing this report.

Introduction

The Expert Consultation on Cotton Pests and their Control in the Near East Region was held from 5 to 9 September 1994 in Izmir, Turkey. The consultation was organized jointly by the FAO Regional Office for the Near East (RNE), Cairo, and the FAO Plant Protection Service (AGPP) and hosted by the Plant Protection Research Institute at Bornova-Izmir, Turkey.

The objectives of the consultation were to assess cotton pest problems in the Near East and the impact of present control measures and to recommend sound strategies for the development and implementation of an effective, sustainable, economic and environmentally friendly integrated cotton pest management programme.

Technical officers from Egypt, Morocco, Pakistan, the Sudan, Turkey and Yemen participated in the consultation. In addition reports were received from the Islamic Republic of Iran, Iraq and the Syrian Arab Republic. Invited speakers from France and the United States also took part in the meeting together with the FAO Senior Regional Crop Protection Officer for Africa and the Secretary of the International Integrated Pest Management (IPM) Working Group. The programme of the consultation and the list of participants are given on p. 269-273, together with a proposal for a regional cotton integrated pest management project for the Near East.

OPENING SESSION

Mr Coskun Saydam, Director of the Plant Protection Research Institute, Bornova-Izmir, welcomed the participants to the expert consultation. He emphasized the importance of cotton to the agricultural economies of the countries in the Near East Region. He pointed out that, although several different varieties of cotton were grown in the various countries, the pest problems encountered were common to all. He expressed the hope that the consultation would develop suitable IPM strategies for farmers in the region as well as laying the foundation for increased cooperation among the countries.

Mr Doorenbos, the FAO Representative in Turkey, welcomed the participants to the consultation on behalf of Dr Jacques Diouf, Director-General of FAO. He referred to Agenda 21 of the United Nations Conference on Environment and Development (UNCED) which specifically called for the implementation of IPM by governments. He indicated that FAO was actively promoting IPM in Member Countries as a preferred strategy for plant protection. This meant the integration of the contributions of the many sectors involved: governments, international organizations, donors, non-governmental organizations (NGOs), researchers covering different disciplines, the private sector and, most important, the farmers and their organizations. Plant pest control crosses international borders and regional cooperation is essential. He emphasized the need for sustainable agricultural production in the region. Mr Doorenbos said that the participants in the consultation represented the best expertise existing in the region and that they should take advantage of the meeting to address the needs of farmers in implementing successful IPM programmes.

Mr Mahmoud Taher, FAO Regional Plant Protection Officer for the Near East, on behalf of Dr Atef Bukhari, FAO Assistant Director-General and Regional Representative for the Near East, welcomed the participants to the expert consultation. He expressed sincere thanks and appreciation to the Government of Turkey, the Ministry of Agriculture and Rural Affairs and the Plant Protection Research Institute in Bornova-Izmir, for having agreed to host the consultation and for providing all the support needed for its organization and success.

He said that, within the Near East region, cotton is a major cash crop in Afghanistan, Egypt, the Islamic Republic of Iran, the Syrian Arab Republic, Iraq, Morocco, Pakistan, Somalia, the Sudan, Turkey and Yemen. Lint is used domestically for textile manufacture and cottonseed is processed for cooking oil and animal feed. The growing and processing of cotton provides employment for large numbers of people. In addition, the export of raw lint and processed goods is a major source of hard currency. Statistics on cotton production in the FAO Near East Region indicate that cotton is grown on about 4.5 million ha, representing about 12 percent of the world cotton area.

Because of the great economic importance of cotton, catastrophic losses in yield from pests can result in significant financial shortfalls in affected countries. For this reason, governments go to great lengths to protect yields from pests, providing technical assistance, credits and pesticide subsidies to cotton farmers. In spite of these measures, pests still cause significant yield and quality losses every year. When losses to pests are combined with the costs of chemical treatments, the total cost amounts to hundreds of millions of US dollars each year. The introduction of chemical pesticides into the cotton production system was one of the essential factors contributing to increased yield. However, the use of chemicals has created many new pest problems, some of which are now more serious than the original target pests for which the pesticides were intended. These problems emphasize the need to develop IPM programmes for cotton that are less dependent on pesticides, are environmentally friendly and are sustainable.

Mr Taher indicated that the consultation was organized with the objective of assessing the cotton pest situation in the FAO Near East Region; evaluating the impact of present control measures and developing appropriate strategies to help Near Eastern farmers in the implementation of integrated pest control practices that are effective, environmentally safe, economical and sustainable.

Mr Sebastian Barbosa, Senior Officer (IPM), Plant Protection Service, FAO, Rome, emphasized the importance of cotton growing for the region and stressed the need to develop IPM programmes for this crop. Reduction in the use of pesticides on cotton would bring enormous benefits to human health and the environment, while increasing farmers' profits and reintroducing sustainability to the ancient cropping systems of the region. Mr Barbosa acknowledged the existence of many technologies in the region that could be incorporated immediately into current plant protection activities. He insisted that the IPM constituency in the region should be broadened to include extension agents and farmers, indicating that to date research on IPM had been fragmented, satisfying only the personal interests of entomologists, weed scientists and plant pathologists, without necessarily solving the farmers' problems.

In his address Mr Yuksel Ayhan, Governor of Bornova District, Izmir, welcomed the participants and indicated that, despite technological advancements, synthetic fibres were not considered desirable alternatives to natural fibres. Therefore cotton would continue to retain its importance in agricultural production as well as its socio-economic status in producing countries. Mr Ayhan stated that, since the Second World War, pesticides had been considered the best solution to cotton pest problems. However, their proven adverse impact on production, quality and the environment had necessitated the search for better control tactics that were effective, safe and sustainable. He concluded that serious efforts should be made to develop and implement a cotton IPM programme to ensure the sustainability of the crop and the safety of the environment.

Mr Muammer Yasarbas, Assistant General Director, General Directorate of Protection and Control, Ministry of Agriculture and Rural Affairs, Ankara, also welcomed the participants. He commented on the socio-economic importance of cotton throughout the world and indicated that Turkey is one of the outstanding cotton producers and that the crop has a very important place in the national economy. However, cotton in Turkey is threatened by 40 pest species known to cause considerable economic losses.

Turkey is well aware of the importance of IPM for sustaining agriculture, maintaining biological diversity and protecting the environment. It initiated its first IPM project in cotton in 1970. As a result, pesticide application has been reduced from eight to four sprays per season. This project was followed by a series of IPM projects in cotton and other crops. Mr Yasarbas concluded that Turkish experts would be very pleased to exchange and share knowledge and experiences with those in other Near East countries.

KEYNOTE ADDRESS
Cotton production and protection with special reference to the mediterranean-Near East region
Michel Braud

Mr Braud gave an overview of the major problems faced by the cotton industry in recent years and the challenges to be confronted in the development of institutions and individuals to make cotton production profitable and sustainable. He referred to the major changes that have taken place recently in the former Soviet Union and led to a considerable reduction in the area devoted to cotton cultivation. He also referred to Egypt and the Sudan, which have reduced their traditional areas under cotton, in contrast to Greece, Turkey and the Syrian Arab Republic, where cultivation has been increased. Mr Braud considered major cotton production constraints in the region including high latitude (short growing season), land fragmentation (small farmholdings), low rainfall and shortage of irrigation water and the presence of some of the world's most important cotton pests (insects, diseases and weeds). He referred to the major pests in each country and indicated the yield losses caused by each of them. He praised the quality of the cotton produced in the region, which despite the many constraints compared positively with other major cotton-producing areas of the world. He emphasized the potential of the region to regain its position in the world cotton markets.

Finally, Mr Braud briefed the meeting on the purpose and activities of the cotton research network, highlighting its interregional mandate which covers Europe, the Maghreb, the Near East and the cotton-producing republics of the Commonwealth of Independent States. He praised the many activities of the network, indicating its paramount importance in facilitating the exchange of information among cotton researchers of this vast and diverse region.

Part I

Country Reports

Egypt

Gallal Moawad

INTRODUCTION

The cotton sector remains of major importance to the Egyptian economy; it provides income for farmers, supplies the local textile industry, provides employment and earns foreign exchange. According to the agricultural census of 1980 there were just over 1 million cotton holdings, with an area of 0.5 million ha (1.2 million feddans – 1 feddan equals 0.42 ha). Over the past two decades there have been declines in area and production; in the early 1970s these amounted to 0.63 million ha and 1.4 million tonnes of seed cotton. By the late 1980s the area had declined to about 0.4 million ha and production was 0.83 million tonnes of seed cotton. Yield levels remained much the same, at about 2 000 kg of seed cotton per hectare, throughout this period, although they did reach a peak of 2 700 kg per hectare in 1982.

Exports of raw cotton have also declined and now only some 25 percent of the crop is exported, the rest being used locally. This decline in export volume, by 50 percent between 1982 and 1988, was not matched by a similar decline in value, which dropped by only 20 percent, a reflection on the price premium paid for extra-long-staple cottons *(Gossypium barbadense)* from Egypt.

Over the past few seasons, however, there has been increasing dissatisfaction with the quality of Egyptian cotton in its main markets, especially in Japan and Italy. This, together with the high prices demanded by cotton exporters, has caused buyers to look elsewhere for supplies. Unless these quality problems can be solved, especially those arising in postharvest processing and handling, there must be some doubt as to the sustainability of demand for the Egyptian crop in world markets.

CULTURAL PRACTICES
Crop management
Sowing early, by 31 March, avoids much early pest and disease attack and brings the crop to peak flowering and boll formation when temperatures are most favourable, between July and September.

Cottonseed is usually dressed with fungicides to give protection from seed-borne and seedling diseases. Fungicides used include pencyuron 25% DS or tolchlofos-methyl 20% plus thiram 30% DS, both applied at the rate of 3 g per kilogram seed. Carboxin 35% plus captan 35% DS is also used, applied at the rate of 3.5 g per kilogram seed.

Research results show that, for maximum yield, cotton should be sown in hills at 15 cm intrarow spacing and 20 to 25 cm interrow spacing.

Fertilizer
Cotton in Egypt shows a response to nitrogen (N) – the amount of nitrogen required depends on the previous use of the land, less being needed following fallow periods than after wheat or rice. Rates of application vary from 76 to 107 kg N per hectare, being higher in upper Egypt, where the medium-long-staple cottons are grown, than in the delta where the extra-long-staple cottons are liable to lodge under heavy nitrogen application. On some soils cotton also shows a response to phosphate fertilizers and relatively small quantities (200 to 500 kg superphosphate per hectare) may be used. Potassium fertilizers are not needed on Egyptian soils.

Irrigation
All Egyptian cotton is grown entirely under irrigation and between nine and 12 irrigations are required to bring the crop to maturity. Irrigation is stopped in time to allow the crop to dry off before the first picking.

VARIETIES
The main commercial varieties grown in Egypt can be classified into three categories: extra-long-staple, long-staple and medium-long-staple. The first group is grown mainly in the northwest of the delta region and includes Giza 45, Giza 70, Giza 76 and Giza 77. The long-staple varieties Giza 75 and

Giza 81 are grown in the remaining delta areas, the Fayoum and north of Beni Suef. The medium-long-staple variety Dandarah is grown in the north of Beni Suef, Asyut and Sohag, while Giza 80, another medium-long-staple variety, is grown in Minya and the southern districts of Beni Suef governorate.

PESTS

Insects and mites

Insect and mite attacks on cotton in Egypt occur in three fairly distinct phases:

- early season (cutworms, thrips, aphid);
- mid-season (leafworm, American bollworm, aphid);
- late season (pink bollworm, spiny bollworm, whitefly).

Pink bollworm and leafworm are considered the key pests, although in the last decade aphid has become important and whitefly could become a problem.

Agrotis *spp. (Lepidoptera: Noctuidae).* The greasy cutworm, *Agrotis ipsilon* (Hfn.), is the most common of several species of cutworm occurring in Egypt. The pest cuts cotton seedlings down at soil level. Damage is worst on fields that, for various reasons, receive inadequate preparation for cotton after the previous crop (often berseem [lucerne/alfalfa], *Medicago sativa* L.). It is important that the soil should be allowed to dry out completely. When infestations are high chemical control may be necessary, either by means of poison baits or by spraying the cotton seedlings. In the 1991 season about 10 000 ha were treated for cutworm, but in some years over 30 000 ha may need to be treated.

Thrips tabaci *(Lindeman) (Thysanoptera: Thripidae).* Thrips attacks the leaves and terminal buds of cotton seedlings and severely infested plants may be stunted or die, reducing the stand. Infestations may be patchy or cover a whole field. Early-sown cotton is less liable to attack and cotton fertilized and irrigated according to recommendations is able to survive an infestation.

The economic threshold is eight to 12 thrips per seedling, depending on the age of the plant. The use of economic thresholds to determine the need for insecticide applications has reduced considerably the area sprayed against thrips in recent years which now averages between 5 and 10 percent of the total cotton area.

Aphis gossypii *(Glov.) (Hemiptera: Aphididae).* Aphid attacks young cotton in April and May and is becoming increasingly important late in the season. In 1991 about 10 percent of the cotton area was sprayed to control aphid.

Tetrancychus *spp. (Acari: Tetranychidae).* Red spider mite became important on cotton in the 1950s following the introduction of organochlorine insecticides. When these were largely replaced by organophosphorus insecticides in the 1960s, the importance of the mite declined. In most years it is necessary to treat only 6 000 to 25 000 ha against mites and in 1984 less than 2 000 ha were treated. A mixture of dicofol and dimethoate (Kelthane S) is marketed to control the complex of early-season sucking pests, including mites.

Spodoptera littoralis *(Boisd.) (Lepidoptera: Noctuidae).* The Egyptian cotton leafworm has in the past been the most serious pest of cotton in Egypt and it remains, potentially at least, a key pest. It is extremely polyphagous, feeding on a wide range of field and vegetable crops, but in crop rotation berseem is the most important host. Berseem provides the main alternate host during the winter between cotton crops and it is berseem that provides the main source of infestations on cotton.

Egg-laying on cotton begins in May and the first generation of larvae reaches peak numbers in June. Second and third generations follow in July and August. Eggs are laid in masses of 200 to 800 on the lower surface of the leaf. Up to 50 000 egg masses per hectare may occur, but populations vary enormously from year to year and according to location. Eggs hatch within three days of oviposition. The area sprayed with insecticides against leafworm varies from 10 to 65 percent of the total area; in 1991 some 69 000 ha were treated.

Spodoptera exigua *(Hb.) (Lepidoptera: Noctuidae).* The lesser, or beet, armyworm is similar in its life history and feeding habits to the Egyptian cotton leafworm, but is only a very minor pest in Egypt, occurring early in the season.

Helicoverpa armigera *(Hb.) (Lepidoptera: Noctuidae)*. American bollworm is now not considered a serious pest of cotton in Egypt, but for a period during the 1970s it did develop into a major threat to production and caused considerable concern to the authorities and the cotton farmers. An economic threshold level (ETL) of five to ten eggs or larvae per 100 plants was set, but this was based on ETLs from other countries and not evaluated in Egypt. Using this ETL an area of 126 000 ha (25 percent of the cotton area) was sprayed against American bollworm in 1975. Intensive research established that the increased importance of American bollworm was probably caused by inappropriate choices of insecticides to control other pests in June and July and that natural enemies were very important in regulating American bollworm numbers (populations could be lower on unsprayed plots than on sprayed plots). On the basis of this research the ETL was raised to 20 small larvae per 100 plants and the area needing to be treated declined to between 500 and 8 000 ha between 1976 and 1979.

Pectinophora gossypiella *(Saund.) (Lepidoptera: Gelichiidae)*. Pink bollworm is the principal pest of cotton in Egypt, as indicated by the fact that 84 percent of insecticide use on cotton is directed against this pest. Pink bollworm overwinters as a diapause larva in the bolls remaining on stalks stored for fuel. The main period of infestation on cotton occurs between July and September. Ministry of Agriculture technicians scout cotton fields regularly from July onwards for pink bollworm. One hundred boll samples are collected from fields, especially those near villages where infestations begin first, and are subsequently examined for the presence of pink bollworm larvae. The ETL is 5 to 10 percent infestation of green bolls. If insecticides are needed for control they are applied at 15- to 21-day intervals, a total of three to four applications usually being sufficient. Yield losses vary, from 4-7 percent in years of light infestation up to 30 percent if cotton is not treated.

Earias insulana *(Boisd.) (Lepidoptera: Noctuidae).* Spiny bollworm is most significant as a pest in the more southerly cotton-growing areas, where it may be of equal importance to pink bollworm.

Bemisia tabaci *(Gen.) (Hemiptera: Aleyrodidae).* Whitefly increased in importance for a period some ten years ago but is usually only of minor significance at present. Experience in cotton-growing countries elsewhere, however, indicates the threat this pest could pose to production if pest management techniques are not properly applied. Natural control, especially by parasitoids such as *Encarsia* sp. and *Eretmocerus* sp. (Hymenoptera: Aphelinidae), is the main factor regulating population numbers. Inappropriate selection of insecticides could easily lead to the re-emergence of whitefly problems in Egypt.

Weeds
The following are the main weed competitors of cotton in Egypt. (Broad-leaved weeds predominate.)

Broad-leaved weeds. *Portulaca oleracea* L. (Portulacaceae); *Hibiscus trionum* L.(Malvaceae); *Malva parviflora* L. (Malvaceae); *Corchorus olitorius* L. (Tiliaceae); *Xanthium spinosum* L. (Compositae); *Chenopodium album* L. (Chenopodiaceae); *Amaranthus retroflexus* L. (Amaranthaceae); and *Euphorbia heterophylla* L. (Euphorbiaceae).

Annual grass weeds. These include: *Echinochloa colona* (L.) Link (Gramineae); and *E. crus-galli* (L.) P. Beauv. (Gramineae).

Perennial grass and sedge weeds. *Cynodon dactylon* (L.) Pers. (Gramineae); and *Cyperus* spp. (Cyperaceae).

Weed control is mainly by hand-hoeing at 21, 35, 50 and 60 days after sowing. Where pre-emergence herbicides are used an additional hoeing may be required after seven weeks. The main herbicides used in cotton are shown in Table 1.

TABLE 1
Herbicides used in Egypt

Active ingredient (percentage and formulation)	Target weeds
Fluometuron 80% wp	Winter annuals
Pendimethalin 50% ec + Fluometuron 80% wp	Winter or summer annuals
Butralin 48% ec + Fluometuron 80% wp	Winter or summer annuals, pre-emergence
Pendimethalin 50% ec	Summer annuals, pre-emergence
Butralin 48% ec	Summer annuals, pre-emergence
Fluazifop-p-butyl 12.5% ec	Annual grasses, postemergence
Sethoxydim 20% ec	Annual grasses, postemergence
Quizalofop 10% ec	Annual grasses, postemergence
Fluazifop-p-butyl 12.5% ec	Perennial grasses, postemergence
Sethoxydim 20% ec	Perennial grasses, postemergence
Quizalofop 10% ec	Perennial grasses, postemergence
Haloxyfop 12.5% ec	Perennial grasses, postemergence

CONTROL METHODS
Chemical control

Pest resistance to insecticides is a problem in Egypt and, in an attempt to manage the problem, insecticide groups are alternated in the spraying sequence. Normally three to four spray applications are required each season. For the first application an organophosphorus insecticide is used to control early sucking pests or, where this application is directed mainly against residual leafworm populations or early bollworm attack, a carbamate insecticide combined with an insect growth regulator is used. For the second application, usually against bollworms, a pyrethroid is used. Later applications revert to organophosphates or carbamates. Applications are made on the basis of ETLs established by regular pest monitoring by staff of the Ministry of Agriculture. Table 2 lists the principal insecticides used on cotton.

Pesticides for use in Egypt are first tested over three consecutive seasons at more than ten experimental stations belonging to the Ministry of Agriculture, universities and research organizations. Efficacy against target

TABLE 2

Principal insecticides used on cotton in Egypt

Active ingredient (percentage and formulations)	Target pest
Methamidophos 60% ec Furathiocarb 40% ec Omethoate 80% sl Monocrotophos 40% sl Phosalone 35% ec Cyanophos 50% ec	Aphid (seedling stage)
Methamidophos 60% ec Furathiocarb 40% ec Omethoate 80% sl Monocrotophos 40% sl Dicofol 18.5% ec + sulphur Dicofol 18.5% ec + fenitrothion 20% ec	Thrips (seedling stage)
Binapacryl 40% ec Dicofol 18.5% ec	Red spider (seedling stage)
Monocrotophos 40% sl Triazophos 40% ec Cyanophos 50% ec	Cutworms
Methamidophos 60% ec Dicofol 18.5% ec Dicofol 18.5% ec + fenitrothion 20% ec	Mole cricket
Methamidophos 30% ec + triflumuron 3% ec Profenofos 72% ec	Leafworms
Deltamethrin 2.5% ec Fenvalerate 20% ec Cypermethrin 10% ec Permethrin 5% ec Fenpropathrin 30% ec Alpha-cypermethrin 25% ec Esfenvalerate 50% ec Sulprofos 52% ec Chlorpyrifos 48% ec Triazophos 40% ec Thiodicarb 37% sc Carbaryl 85% wp Profenofos 72% ec Methamidophos 60% ec	Bollworms

Note: For each active ingredient there are usually several formulations and, often, combinations under a variety of trade names. Dosages vary accordingly.

pests, effects on non-target organisms and the environment generally and hazards to humans and livestock are all assessed. The results of this evaluation programme are considered by the government committee responsible for pesticide registration and regulation.

Resistance has resulted in a number of insecticides becoming ineffective and these have been phased out. From 1955, toxaphene was used for leafworm and bollworm control, but in 1961 it failed to give control and cotton production was reduced by one-third. Carbaryl, trichlorfon, monocrotophos and other insecticides have been dropped from the list of chemicals used against leafworm because of resistance. Each year resistance levels in populations of the major pests are monitored twice, before and after the cotton season, in all the cotton-growing governorates. The results are used by the Ministry of Agriculture to determine pesticide recommendations for the following season and the requirements of the different governorates. The cost of insecticides over a season averages about 143.00 Egyptian pounds (£E) (US$42) per hectare, but with a government subsidy the farmer pays only £E48.00 (US$14) per hectare.

Both ground and aerial application methods are used for insecticides and pheromones, although aerial application is in decline. Volume rates of application for conventional insecticides vary from 24 litres per hectare for aerial application to 1 500 litres per hectare with motorized ground sprayers.

The organization of aerial spraying is the responsibility of the Ministry of Agriculture's Agricultural Aviation Department, which is also responsible for locust control. The financial and technical aspects of cotton pest control are handled centrally, while operational matters are normally dealt with at governorate level by department inspectors and engineers. The actual aerial spraying is done by private-sector contractors. The system is rather inflexible, with decisions being made in advance of the season rather than in response to pest problems as they arise.

Legislative control
The carryover of cotton leafworm from berseem to cotton is discouraged by legally backed requirements to cease irrigation of berseem by 10 May and to include kerosene, at the rate of 70 litres per hectare, in the final irrigation water. These measures ensure that the berseem crop terminates early and loses its attraction for egg-laying by the leafworm and that the soil is too dry and unattractive to permit successful pupation.

Biological control
The role of natural enemies in controlling cotton pests is becoming

increasingly recognized and the pest management strategy is now designed to enhance this role. Insecticide applications are not permitted on cotton between May and the first week of July to allow populations of parasitoids and predators to become established. During this period, leafworm has traditionally been controlled by the collection of egg masses by children, although this is now practised less frequently, unless there is a major outbreak, because it was found that sufficient control could usually be exercised by natural enemies.

Cultural control

Cotton grown under high standards of husbandry, according to recommendations for land preparation, fertilizer use, irrigation practices, weed control, destruction of crop residues, rotational practices and appropriate choice of variety, is better able to withstand pest attack or avoid it altogether.

NEW DEVELOPMENTS IN COTTON PEST MANAGEMENT
Pheromones

The use of pheromones to control pink bollworm by the mating disruption technique is now well established in Egypt. In 1994, 50 percent of the cotton area was treated with pheromones. Four to five applications of pheromones are made from early in the season with the aim of preventing population buildup. The technique is a prophylactic one and ETLs are not employed. If necessary, pheromones are supplemented by applications of conventional insecticides late in the season. A major advantage of pheromones is that they do not affect populations of beneficial insects and they may completely replace conventional insecticides for pink bollworm control in the near future.

The use of pheromones to control the other major pests, i.e. leafworm and spiny bollworm, is still at the experimental stage, as are other methods, including the use of pheromones for mass trapping, lure and kill techniques and population monitoring.

Viruses

A major programme to investigate the use of naturally occurring nuclear polyhidrosis virus (NPV) to control leafworm has been carried out in Egypt.

Trials have shown that larval control with the virus can be equal or superior to control by conventional insecticides or by hand-collection of egg masses. Viruses have the advantages that they do not contaminate the environment, do not affect the beneficial species and can be produced locally.

Bacillus thuringiensis (Bt) offers a promising alternative to conventional insecticides for the control of many pest species and its use is currently being investigated in Egypt. There is concern that pests will develop resistance to Bt in the same way that they have to conventional chemicals and its use will need to be monitored closely.

IPM STRATEGY

Integrated pest management in Egypt adheres to the following practices:

- The carryover of leafworm from berseem to cotton is discouraged by early cessation (10 May) of berseem irrigation and the use of kerosene in irrigation water.
- Chemical control of early pests, especially thrips, is based on ETLs in order to prevent unnecessary use of insecticides early in the season with adverse effects on natural enemy populations.
- Hand-collection of leafworm egg masses preserves populations of natural enemies which normally give adequate control, even without hand-collection.
- The use of insecticides for leafworm control is not usually necessary until after the natural enemies reach peak populations in early July.
- The use of insecticides for leafworm and bollworm control is based on the use of ETLs, the choice of pesticides with selective action and, where possible, the use of insect growth regulators.
- The overall strategy is to minimize the use of pesticides, especially early in the season.

As a result of the introduction of the IPM approach to pest control, cotton leafworm is of diminishing importance as a pest; early-season sucking pests are also of less importance; yield losses caused by bollworm have declined from about 20 percent in the 1950s to 5-8 percent at present; and the average number of sprays per season has dropped to three or four, thus reducing production costs and reducing environmental impact. Yields continue to fluctuate but the overall trend is upwards. The record yield of 1 190 kg lint per hectare, was achieved in 1982.

INFRASTRUCTURAL SUPPORT FOR COTTON IPM
Government organizations
Plant Protection Research Institute (PPRI). PPRI's mandate is to carry out research on the protection of agricultural crops from attack by insects and other pests, including research on the natural enemies of pests, and to provide recommendations on pest control. Its activities include:
- testing new insecticides against cotton pests;
- evaluating alternatives to chemical pesticides, including pheromones, viruses and *Bacillus thuringiensis*;
- control of vegetable and orchard pests;
- assessing the role of natural enemies in pest management;
- preparation of insects and mites for reference collections;
- research on plant diseases.

PPRI suffers from a lack of resources, including funding for new equipment, maintenance and running costs, and needs overseas training fellowships for staff at postgraduate level.

Plant protection services. The Ministry of Agriculture is responsible for plant protection services, including pest monitoring, supervision of control operations and the enforcement of control legislation.

Extension services. The extension services collaborate with plant protection specialists from the research institutes and the plant protection services in running training courses and disseminating crop protection information in different ways.

KEY PESTS
The actual and potential key pests of cotton in Egypt are: Egyptian cotton leafworm *(Spodoptera littoralis)*, pink bollworm *(Pectinophora gossypiella)*, aphid *(Aphis gossypii)* and whitefly *(Bemisia tabaci)*.

KEY PERSONNEL INVOLVED IN COTTON PEST CONTROL IN EGYPT

First Under-Secretary of State for Extension Sector, Ministry of Agriculture and Land Reclamation (MOA).

Under-Secretary of State for Pest Control (MOA).

Director, Plant Protection Research Institute (PPRI), Agricultural Research Centre (ARC), (MOA).

General Director of Pest Control, (MOA).

Director, Plant Pathology Research Institute, (ARC), (MOA).

Head, Cotton Leafworm Research Department, (PPRI), (ARC).

Head, Cotton Bollworm Research Department, (PPRI), (ARC).

Head, Sucking Insect Research Department, (PPRI), (ARC).

Head, Pesticides Evaluation Research Department, (PPRI), (ARC).

Directors of pest control in different governorates and districts, (MOA).

References

Abdeen, S.A.O., Gadallah, A.I., Nagwa M. Hussein & Moawad, G.M. 1986. Some toxicological and biochemical effects of bacterium *Bacillus thuringiensis* on the American bollworm *Heliothis armigera* (Hbn). Faculty of Agriculture, Ain Shams University, Cairo, Egypt, *Annals Agric. Sci.*, 31(2): 1445-1461.

Abd El-Salam, N.M., Amira, M. Rashad, Moawad, G.M. & El-Hamaky, M.A. 1991. Evaluation and initiation of bollworm control on basis of young larval count of bollworm infestation in cotton fields in Egypt. *Bull. Soc. Entomol.*

Abo El-Ala, M. & El-Baradeiy, M. 1958. Results of application of major nutrients on cotton yield. *2nd Cott. Conf. Egypt,* p. 256-266. (in Arabic)

Campion, D.G. & Hosny, M.M. 1982. Recent advances in the use of pheromone in developing countries with particular reference to mass trapping for the control of *Spodoptera littoralis* and mating disruption for the control of *Pectinophora gossypiella. Les mediateurs Chimiques Agissant sur le Comportement des insects,* International Symposium, 16 to 20 November 1981 Versailles, Paris, France, Institute National de la Recherche Agronomique, 72(8): 5501.

Campion, D.G. & Hosny, M.M. 1987. Biological, cultural and selective methods for control of cotton pests in Egypt. *Insect Science and its Application,* 8: 4-6.

Critchley, B.R., Campion, D.G., McVeigh, L.J., McVeigh, E.M., Cavanagh, G.G., Hosny, M.M., Nasr El-Sayed, A., Khidr, A.A. & Naguib, M. 1985. Control of pink bollworm, *Pectinophora gossypiella* (Saunders) (Lepidoptera: Gelechiidae), in Egypt by mating disruption using hollow fibre, laminated-flake and microencapsulated formulations of synthetic pheromone. *Bull. Entomol. Res.,* 75: 329-345.

Critchley, B.R, Campion, D.G., McVeigh, L.J., Hunter-Jones, P., Hall, D.R., Cork, A., Nesbitt, B.F., Marrs, G.J., Jutsum, A.R., Hosny, M.M. & Nasr El-Sayed, A. 1983. Control of pink bollworm *Pectinophora gossypiella* (Saunders) (Lepidoptera: Gelechiidae) in Egypt by mating disruption using an aerially applied microencapsulated pheromone formulation. *Bull. Entomol. Res.,* 73: 289-299.

Downhan, M.C.A., McVeigh, L.J. & Moawad, G.M. 1989. Prospects for the control of Egyptian cotton leafworm *Spodoptera littoralis* (Boisd) using lure and kill strategy. *Proc. 1st. Int. Conf. Econ Entomol.,* Vol. II, p. 107-115.

Eid, M.T. & Hemissa, M.R. 1958. Nitrogenous fertilization for cotton. *2nd Cott. Conf, Egypt.,* p. 372-378. (in Arabic)

El-Adl, M.A., Hosny, M.M. & Campion, D.G. 1988. Mating disruption for the control of pink bollworm *Pectinophora gossypiella* (Saunders) in the delta cotton-growing area of Egypt. *Trop. Pest Manag.,* 34(2): 210-214.

El-Fateh, S.M., El-Hamaky, M.A., Moawad, G.M. & Hussein, Nagwa M. 1988-89. Large-scale evaluation of pheromones in reducing the population density of the pink bollworm, *Pectinophora gossypiella* (Saund). *Bull. Entomol. Soc. Egypt, Econ. Ser.,* 17: 19-28.

El-Hamaky, M.A., Watson, W.A. & El-Fateh, S.M. 1987. Potency of certain insecticides and insect growth regulators and their mixtures on *Spodoptera littoralis* (Boisd.). Tanta University, Tanta, Egypt, *J. Agric. Res.,* 13(4): 1198-1212.

El-Kifi, A.H., Nasr El-Sayed, A. & Moawad, G.M. 1972. Factors affecting longevity and reproductive potentiality of the black cutworm, *Agrotis ipsilon* (Hufn.) moth. *Bull. Entomol. Soc. Egypt,* 56: 195-200.

El-Saadany, G.G., El-Shaarawy, M.F. & El-Refaei, Sh.A. 1975. Determination of the loss in cotton yield as being affected by the pink bollworm *Pectinophora gossypiella* (Saund.) and the spiny bollworm *Earias insulana* (Boisd.). *Z. Ang. Entomol.,* 79(4): 357-360.

El-Sayed, A.N. 1965. Ecological studies on the resting stage of the pink bollworm in Egypt. *Bull. Entomol. Soc. Egypt,* 49: 75-78.

El-Sayed, G.N., El-Guindy, M.A., Moawad, G.M., Madi, S.M. & Farrag, A.M.I. (1978-79). Spatial distribution of resistance to insecticides in different field strains of the greasy cutworm, *Agrotis ipsilon* (Hufn.). *Bull. Entomol. Soc. Egypt, Econ. Ser.,* 12: 71-82.

El-Sayed, G.N., El-Guindy, M.A., Madi, S.M., Dogheim, S.M. & Moawad, G.M., 1980-81. Geographical distribution of organophosphorus insecticide-resistant strains of the cotton leafworm, *Spodoptera littoralis* (Boisd.) in the Nile Delta. *Bull. Entomol. Soc. Egypt, Econ. Ser.,* 12: 71-82.

El-Sayed, M.T. & Abd-El-Rahman, H.A. 1960. On the biology and life history of the pink bollworm *Pectinophora gossypiella* (Saunders). *Bull. Entomol. Soc. Egypt,* 44: 71-90.

El-Sayed, M.T. & Rustom, Z.M.F. 1960. Factors affecting termination of the resting stage of the pink bollworm. *Bull. Entomol. Soc. Egypt,* 44: 265-282.

El-Shaarawy, M.F., El-Saadany, G. & El-Rafaei, Sh.A. 1975. The economic threshold of infestation for the cotton bollworms and yield in Egypt. *Z. Ang. Entomol.,* 79(3): 276-281.

Emara, S.A, Moawad, G.M. & Gadallah, A.I. 1991. Effect of three formulations of *Bacillus thuringiensis* on *Helicoverpa armigera* (Hb.) *4th Arab Congress of Plant Protection* (Cairo, Egypt, 1 to 5 December 1991), Vol. II. p. 279-287.

Farrag, S.M. 1976. Studies on the natural mortality of *P. gossypiella* (Saund.) with special reference to its bacterial diseases. Faculty of Agriculture, Cairo University, Cairo, Egypt. (M.Sc. thesis)

Fayad, Y.H. & Ibrahim A.A. 1980. Effect of some new insecticides of cotton leafworm on the number of predators in cotton fields. *Proc. 1st Conf. Plant Prot. Res. Inst., Cairo, Egypt,* 11: 33-384.

Hosny, M.M. & Metwally, A.G. 1974. Stored cotton sticks as a source of pink bollworm infestation. *Bull. Entomol. Soc. Egypt,* 58: 153-161.

Hosny, M.M., Nasr El-Sayed, A. & Shafei, S.M. 1978. Catches of *S. littoralis* (Boisd.) male moths in light and pheromone traps in the delta and Middle Egypt. *4th Conference on Pest Control,* Cairo, Egypt, NRC.

Hosny, M.M., El-Saadany, G., Moawad, G.M. & Hossain, A.M. 1991. The effect of biotic factors on population size of *Pectinophora gossypiella* (Saunders) diapause larvae. *4th Arab Congress of Plant Protection,* Cairo, Egypt, 1 to 5 December 1991, Vol. II, p. 301-308.

Hosny, M.M., El-Sayed, G., Moawad, G.M. & Topper, C.P. 1983. Evaluation of the efficacy of certain bacterial insecticides (*Bacillus thuringiensis*) in controlling *Spodoptera littoralis* in Egypt. *Agric. Res. Rev.,* 61: 45-55.

Hosny, M.M, Moawad, G.M., El-Saadany, G. & Hossain, A.H. 1991. Effect of crop rotation on the overwintering population of the pink bollworm. *4th Arab Congress of Plant Protection,* Cairo, Eygpt, 1 to 5 December Vol. II, pp. 10-13.

Hosny, M.M., Topper, C.P., Moawad, G.M. & El-Saadany, G. 1984. The economic damage thresholds of *Spodoptera littoralis* (Boisd.) (Lepidoptera: Noctuidae) on cotton in Egypt. *Agric. Res. Rev.,* 62(1): 1-10.

Hosny, M.M., El-Saadany, G., Iss-Hak, R., Nagar, S., Nasr, E., Moawad, G.M., Critchley, B.R., Topper, C.P. & Campion, D.G. 1980. Report on the use of pheromone-baited traps for the control of the Egyptian cotton leafworm and the bollworm by mass trapping on cotton in Egypt. *16th International Congress of Entomology,* Kyoto, Japan.

Hosny, M.M., El-Saadany, G., Iss-Hak, R., Nasr El-Sayed, A., Naguib, M., Moawad, G.M., Campion, D.G., Critchley, B.R., Mckinley, D.J., McVeigh, L.J. & Topper, C.P. 1983. Techniques for the control of cotton pests in Egypt to reduce the reliance on chemical pesticides. *International Conference on Environmental Hazards of Agrochemicals in Developing Countries in Collaboration with Man and the Biosphere (MAB) Programme,* p. 103-104. University of Alexandria, Alexandria, Egypt, UNEP.

Kamal, M. 1936. Recent advances in the control of pink bollworm by natural enemies. *Bull. Entomol. Soc. Egypt,* 20: 259-293.

Khidr, A., Moawad, G.M. & McVeigh, L.J. 1984. Sex pheromones of Lepidoptera: the effect of control measures using pheromones and insecticides in cotton fields. *Al-Azhar J. Agric. Res.,* 1: 106-112.

McKinley, D.J., Moawad, G.M., Jones, K.A., Grzywacz, D. & Turner, C.

1989. The development of nuclear polyhedrosis virus for the control of *Spodoptera littoralis* (Boisd.) in cotton. *In:* M.B. Green & D.J. de B. Lyon, eds. *Pest management in cotton,* p. 93-100. Chichester, UK, Ellis Horwood.

Metwally, A.G., Hafez, A.A.E. & El-Bishry, H. (1976). The use of pheromone traps as a survey tool for *Pectinophora gossypiella* (Saund.). *Agric. Res. Rev.,* 54: 9-15.

Moawad, G.M. 1974. Effect of certain species of bollworms, *P. gossypiella* (Saund.) and *Earias insulana* Boisd., on cotton plant and yield in relation to agricultural practices. Faculty of Agriculture, Al-Azhar University, Egypt. 192 pp. (unpublished Ph.D. thesis)

Moawad, G.M. 1981. Survival of pink bollworm, *P. gossypiella* (Saund.) under various soils and climatic conditions. *Agric. Res. Rev.,* 62(1): 39-43.

Moawad, G.M. 1984. Persistance of *Spodoptera littoralis* nuclear polyhedrosis virus on the cotton plants. *Agric. Res. Rev.,* 62(1): 39-43.

Moawad, G.M. & Abdeen, S.A.O. 1991. Effects of *Heliothis* nuclear polyhedrosis virus on the American bollworm, *H. armigera* (Hbn.). *Bull. Entomol. Soc. Egypt,* 14: 385-95.

Moawad, G.M., Abdel-Halim, S.M., Mckinley, D.J., Grzywacz, D. & Jones, K.A. 1989. Development of nuclear polyhedrosis virus for control of *Spodoptera littoralis* in Egypt. *Proc. 1st. Int. Conf. Econ. Entomol.,* vol. 11, p. 1-9.

Moawad, G.M., El-Saeudy, A.A., Rashad, A.M., Shalaby, M.A.M., and Gadallah, A.I. 1990. Suppression of pink bollworm, *Pectinophora gossypiella* (Saund.) infestation in cotton fields treated with sex pheromones. 3rd Conf. Agric. Dev. Res., Faculty of Agriculture, Ain Shams University, Cairo, Egypt, 22-24 December 1990. *Annals Agric. Sci.,* Special Issue: 531-543.

Moawad, G.M., Hosny, M.M., El-Saadany, G. & Topper, C.P. 1984. Assessment of damage to cotton plants as an indication of the efficacy of nuclear polyhedrosis virus in controlling the cotton leafworm infestation *(Spodoptera littoralis)* in Egypt. *Annals Agric. Sc.,* (Moshtohor), 1: 1087-1093.

Moawad, G.M., Hosny, M.M., El-Saadany, G. & Hossain, A.M. 1991a. Mating disruption as a method of control for the pink bollworm, *Pectinophora gossypiella* (Saund.) in middle Egypt. *4th Arab Congress of Plant Protection,* Cairo, Egypt, 1 to 5 December 1991, vol. II, p. 493-501.

Moawad, G.M., Hosny, M.M., El-Saadany, G. & Hossain, A.M. 1991b. The use of pheromone trap catch data in estimating pink bollworm infestation. *4th Arab Congress of Plant Protection,* Cairo, Egypt, 1 to 5 December 1991, vol. II, p. 490-492.

Moawad, G.M. & Hussein, Abdel-Hamid M. 1980. Effect of certain agricultural practices on the population density of the overwintering pink bollworm, *Pectinophora gossypiella* larvae. *Agric. Rev.,* 1: 265-275.

Moawad, G.M., Khidr, A.A., Zaki, M., Critchley, B.R., McVeigh, L.J. & Campion, D.G. 1991. Large-scale use of hollow fibre and microenacapsulated pink bollworm pheromone formulations integrated with conventional insecticides for the control of the cotton pest complex in Egypt. *Trop. Pest Manag.,* 37(1): 10-15.

Moawad, G.M. & Nasr El-Sayed, A. 1982. The interrelation of spacing, irrigation and fertilization of cotton plants and the number of cotton leafworm egg masses. *Agric. Res. Rev.,* 1: 237-242.

Moawad, G.M., Shalaby, F.F., Metwally, A.G. & El-Gemeiy, H.M. 1982-83. Laboratory pathogencity tests with two commercial preparations of *B. thuringiensis* (Berl.) on the first-instar larvae of the spiny bollworm. *Bull. Soc. Entomol. Egypt,* 64: 137-144.

Moawad, G.M., Topper, C.P., El-Husseini, M. & Kamel, A. 1984. Natural control of the cotton leafworm, *Spodoptera littoralis* egg masses and young larvae in Egypt. *Agric. Res. Rev.,* 62(1): 13-20.

Moawad, G.M., Topper, C.P., El-Husseini, M. & Kamel, A. 1985. Seasonal abundance of parasites and predators of *Spodoptera littoralis* in clover fields in Egypt. *Agric. Res. Rev.,* 63(1): 45-54.

Nasr El-Sayed, A., Badr, N.A., Hamed, M.A. & Ahmed, M.A. 1982. Population density and flight activity of the adult stage of the spiny bollworm, *Earias insulana* (Boisd.). *Agric. Res. Rev.,* 58(1): 167-197.

Nasr El-Sayed, A. & Moawad, G.M. (1972). Some ecological aspects concerning the black cutworm, *Agrotis ipsilon* (Hufn.). *Bull. Soc. Entomol. Egypt,* 56: 139-144.

Salluma, B.M. 1962. Study on the effect of the major elements on cotton fibre. Faculty of Agriculture, Ain Shams University, Cairo, Egypt. (M.Sc. thesis)

Shalaby, F.F., Moawad, G.M., El-Lakawa, F.A. & El-Gemeiy, H.M. 1983.

Laboratory pathogencity tests with a commercial product *Bacillus thuringiensis* (Berl.) against active and resting larvae of the pink bollworm. *Agric. Res. Rev.,* 61(1): 23-41.

Tawfik, M.F.S. & Awadallah, K.T. 1970. The biology of *Pyemotes herfsi* (Oudemans) and its efficiency in the control of the resting larvae of the pink bollworm *Pectinophora gossypiella* (Saunders) in ARE. *Bull. Soc. Entomol. Egypt,* 54: 49-71.

Tawfik, M.F.S. & El-Sherif, I.S. 1977. The status of dead cotton bolls as a source for infestation of cotton plants with *Pectinophora gossypiella* (Lepidoptera: Gelechiidae). *Bull. Soc. Entomol. Egypt,* 58: 191-196.

Topper, C.P., Moawad, G.M., McKinley, D.J., Hosny, M.M., Jones, K., Cooper, J., El-Nagar, S. & El-Sheik, M. 1984. Field trials with a nuclear polyhedrosis virus against *Spodoptera littoralis* on cotton in Egypt. *Trop. Pest Manag.,* 30(4): 372-378.

Watson M., El-Hamaky, M.A. & Guirguis, M.W. 1986. Ovicidal action and latent toxicity of certain chitin synthesis inhibitors and their mixtures with natural oil and wetting agent. Tanta University, Tanta, Egypt, *J. Agric. Res.,* 13(4): 1185-1197.

Islamic Republic of Iran

Ahmad Rassipour

INTRODUCTION

Cotton has been grown in Iran for a very long time; ancient documents show that the wearing of cotton clothing was common very early in the history of the country. Before 1862 only short-staple varieties were grown on a limited area for local use. Exports were rare. In 1924 the Iranian-Russian Cotton Company imported seed of American upland cotton *(Gossypium hirsutum)* which was grown experimentally in Mazandaran province. In 1931 the Ministry of Agriculture initiated variety trials in the Varamin region of Tehran province. The most promising material was then used in a selection and breeding programme which resulted in the release to farmers of a new variety, Filestan. By 1934 many ginneries had been built and experimental plots established and consequently production expanded in the regions most favourable to cotton, such as Mazandaran, Gorgan, Garmsar and Varamin. Production steadily increased and had reached 42 000 tonnes by 1942, at which time there were 13 ginneries. During the Second World War, however, production declined and in 1943 was only 16 000 tonnes.

In 1952 FAO sent a cotton specialist to Iran and a new era began in the history of cotton growing in the country. In the five-year period from 1951 to 1955 production increased from 36 000 tonnes to 70 000 tonnes.

A Cotton Council was formed to encourage and support the development of cotton production and processing. This led to the formation of a Cotton Organization to supervise and control the processing of cotton by private and state ginneries. Samples were graded according to international standards and certificates issued by the organization. Cooperation between the Cotton Organization and research and operational organizations resulted in further increases in the area under cotton and in total cotton production, which by 1974 had reached 237 000 tonnes.

COTTON-GROWING REGIONS

Table 3 shows the main cotton-growing regions of Iran with areas under cotton and total production. Cotton can be grown in soils with a pH in the range 5.5 to 9.0, which occur over most of the cultivated areas in Iran. Besides soil and climate, however, there are other factors that determine whether an area is suitable for cotton production. These include soil fertility and the total quantity of water and number of irrigations needed to grow the crop. Other factors that determine cotton production and yield are pests and diseases.

Socio-economic importance of cotton

Cotton contributes to the national economy, supports the local textile industry which supplies national needs and creates employment.

According to the first five-year plan and its annual targets, in 1992 Iran should have produced about 400 000 tonnes of seed cotton and 120 000 tonnes of lint. Cottonseed oil and protein produced from cottonseed are of great value as food for humans and livestock.

TABLE 3

Main cotton-growing regions of Iran; area and total production for 1993

Region	Area *(ha)*	Production of seed cotton *(tonnes)*
Gorgan	120 000	276 000
Khorasan	31 000	60 450
Mazandaran	28 000	39 200
Azarbaijan, Est.	9 000	18 900
Tehran	8 000	14 400
Fars	8 500	16 575
Semnan	3 000	5 400
Esfahan	2 000	4 375
Kermanshah	1 950	2 925
Bakhtaran	900	1 305
Central	600	1 080
Yazd	300	465
Total	**213 250**	**441 075**

Numerous ginneries, oil extraction plants, textile mills and other small factories, including those making carpets, are dependent on cotton production, as is the export of cotton textiles. Until the late 1970s not all the cotton produced was consumed within the country and raw cotton was an important export. Now, because of increased internal demand and declining production, raw cotton is no longer exported.

The cotton sector provides employment for a large number of people. In the 1991 season, cotton was grown on about 190 000 ha by some 85 000 farmers. Cotton production requires about 90 workdays per hectare over the growing and harvesting season. Some 25 percent of the manufacturing workforce is employed in cotton processing and textile production. The cotton sector contributes 21 percent of the total manufacturing sector and is an important factor in other industries such as transport. If the private sector, government and research are included, it is estimated that at least 2 million people are in some way dependent on cotton for their livelihood.

CULTURAL PRACTICES
Land preparation
Land preparation to obtain the conditions suitable for cotton growth and development is very important. Residues from previous crops, such as cotton, maize or sunflower, are buried using rotary cultivators. Fields are deep-ploughed in autumn and then left until spring. Winter rainfall and snow provide water and break down soil clods. Between 20 and 30 tonnes of farmyard manure per hectare is applied at the time of ploughing and is buried to a depth of 45 cm. In spring hoeing to 20 cm depth destroys weeds and creates a seed-bed. If this is not done the cotton root cannot penetrate the soil and the plant is subsequently unable to withstand the dry conditions and high temperatures of summer. Unless a fine tilth is created, with a soil crumb size of 3 to 6 mm (comparable with the size of the cottonseed), germination and emergence are inhibited.

One of the beneficial effects of good land preparation is the control of weeds and pests. Ploughing and hoeing expose overwintering stages of pests on the soil surface where they are destroyed by predators or the effects of the weather. In the temperate cotton-growing zones autumn is the best time to do this. In hotter and drier regions land preparation is concentrated in the period immediately before sowing to avoid problems of erosion and excess water loss through evaporation.

Irrigation

Cotton cultivation requires adequate water from sowing to harvest to obtain good yields and high quality. It is necessary to irrigate cotton crops in areas where rainfall or soil water reserves are insufficient for cotton growth. The amount of water required depends on the climatic conditions and method of cultivation. In areas where rainfall is in the region of 700 to 800 mm per year, it is not necessary to irrigate cotton. In other areas the rate of evaporation determines the amount of irrigation needed. The total quantity required depends on the growth stage of the crop:

- Prior to sowing, 1 500 m^3 of water per hectare is required for soil preparation and the sowing of seed.
- For germination 20 to 40 m^3 of water per hectare is required.
- During flowering and boll formation cotton requires 60 to 90 m^3 of water per hectare per day for 45 to 60 days.
- During the period of boll maturation cotton is less susceptible to dry conditions and irrigation is usually stopped once the first bolls split open.

Cotton usually requires a total of some 6 000 to 9 000 m^3 of water per hectare to produce its crop. In areas of high temperature and light sandy soils, the water requirement is higher, at 12 000 m^3 of water per hectare. The frequency of irrigation varies according to soil characteristics and ambient temperatures, but usually water is applied once every seven to ten days during the growing season.

Methods of irrigation vary according to local traditions and the experience of farmers. Methods include:

- Furrow irrigation – an effective method of irrigation which reduces production costs and economizes on the use of water. Where tubing is used to bring water to the plants, control of water usage is even better. Weeding, spraying and harvesting are all easier when cotton is grown under this system.
- Basin or flood irrigation – used in areas where cotton is grown on small

pieces of land. More water is required in this method and for this reason it is not recommended.

• Overhead irrigation – a very effective method of irrigation but installation and running costs are high.

Fertilizer

Nitrogen requirements vary from 40 to 100 kg per hectare, being higher in the high temperature regions than in more temperate cotton-growing areas. Phosphorus (P) is normally applied as triple superphosphate or ammonium phosphate containing 40 percent P at the rate of 50 to 70 kg P_2O_5 per hectare. Potassium (K) may also be applied, at the rate of 20 to 50 kg K_2O per hectare; this element increases disease resistance. Phosphorus and potassium can be applied at the time of autumn ploughing in the temperate regions.

COTTON VARIETIES

The improvement of cotton began with the introduction of *Gossypium hirsutum* into Iran. Between 1923 and 1932 a number of *G. hirsutum* varieties were imported and tested and in 1946 varieties were imported from the United States. The basic work on cotton development was done by Iranian scientists in collaboration with experts from FAO. Following the initial research the variety Coker 100 Wilt was chosen for multiplication. Various research institutes in Iran have subsequently bred the following varieties through programmes of crossing and selection.

Sahel. Bred by crossing Coker 100 Wilt with 349 in 1959 at the Cotton Research Institute, Varamin. It is high-yielding with good fibre characteristics and is resistant to verticillium wilt. Sahel was released for general cultivation in 1968 and is recommended for areas infested with verticillium, in particular Mazandaran in the Caspian region.

Varamin. Bred by crossing Coker 100 Wilt with 539 at the Cotton Research Institute, Varamin. It is an early, high-yielding variety with good fibre quality. It is recommended for the central regions, Esfahan and Khorasan, where humidity is high and the cotton is liable to be attacked by whitefly and spiny bollworm.

Hopicala. A cross of XQ6-2 Acala and Hopimoencopi obtained from the United States in 1966. It is an early variety with high-quality fibre and is recommended for the Fars region.

Deltapine 16. Higher-yielding and earlier than Acala, this variety is tolerant of hot, dry conditions and is recommended for the southwestern cotton-growing areas in the Fars region.

Dr Omouki. A long-staple variety originally bred for the Jiroft area. It is no longer grown because of a switch from cotton to vegetables in this area.

Bakhtegan. A verticillium wilt-tolerant variety bred for the Fars region.

Oltan. An early variety with short sympodia giving it a columnar structure suitable for machine harvesting. It is recommended for Moghan and the region to the north of Khorasan.

Pake. A variety recommended for the central areas of Iran.

   The production of new varieties with improved characteristics for yield, quality and resistance to pests and diseases, including any new pathogens which may appear, is continuing in Iran through breeding and selection programmes.

PESTS
Insects
Earias insulana *(Boisd.) (Lepidoptera: Noctuidae).* Spiny bollworm is a major pest of cotton in Iran. Its host range includes other cultivated Malvaceae and weeds in the same family, including *Abutilon avicennae* Gaertn., *Hibiscus trionum* L., *Altheae* spp. and *Malva* spp. When no control measures are carried out, losses in cotton yield may range from 25 to 30 percent, although in years of very heavy infestation they may be as high as 70 percent. Larval feeding in bolls leads to infestation by secondary pathogens resulting in boll rots and complete destruction of the lint.

Helicoverpa armigera *(Hb.) (Lepidoptera: Noctuidae).* American bollworm is a major pest of cotton and other crops, including maize, sorghum, beans, tomatoes, lucerne and cucurbits. In normal years yield losses to American bollworm in the Caspian cotton areas of northern Iran range from 10 to 25 percent, but in outbreak years may reach 50 to 75 percent if control measures are not applied.

Spodoptera exigua *(Hb.) (Lepidoptera: Noctuidae).* Lesser, or beet, armyworm is another important pest of cotton and other crops in Iran.

Spodoptera litura *(F.) (Lepidoptera: Noctuidae).* Cotton leafworm is a polyphagous pest in southern Iran where it is found on both cultivated and wild host plants for nine months of the year. It may cause yield losses in cotton of 10 to 15 percent in a normal year without treatment; in outbreak years if not controlled it can cause almost total loss of yield.

Agrotis segetum *(Schiff.) (Lepidoptera: Noctuidae).* Cutworm, a pest of the seedling stage of cotton and other crops, is distributed over most of the cotton-growing areas of Iran.

Tetranychus urticae *(Koch.) (Acari: Tetranychidae).* The two-spotted spider mite is found on cotton and other host plants throughout Iran. Infestations usually begin at the field margins and gradually spread.

Thrips tabaci *(Lind.) (Thysanoptera: Thripidae).* Tobacco thrips is most prevalent on rain-grown cotton in dry springs when damage to seedlings and young plants may be severe. The effects of thrips infestation are to reduce yield and delay harvest.

Bemisia tabaci *(Genn.) (Hemiptera: Aleyrodidae).* Cotton whitefly is a pest of long standing on cotton in Iran but has become increasingly important in recent years, especially in Garmsar in the northern areas and in Fars and Jiroft in the south. Its centre of origin on cotton may have been the coastal areas of the Persian Gulf and the adjacent provinces of Khozustan, Fars, Kerman and

Khorasan. A second species, *Trialeurodes vaporariorum* (Haldeman), has recently become important on cotton in the northern cotton-growing areas of Gorgan and Gonbad, formerly it was mainly a pest of ornamentals. This species, apparently the greenhouse whitefly, may be new to the region. Infestations have required insecticide applications to control them.

So far about 46 host plants of whitefly have been identified in Iran, including crops and ornamentals, trees and weeds.Adults move on to cotton in July from alternate host plants, mainly weeds, on which they appear in June. Whitefly is most important in the Mazandaran region and in Fars. The increase in importance of this pest is partly the result of a failure to observe various cultural practices, including removing host plants from in and around cotton fields. Whitefly has a patchy distribution in the early stage of the development of the crop. In late summer, if control measures are not taken, the infestation begins to expand in area and numbers, and in autumn it moves on to late-sown cotton. Honeydew production is the most important consequence of whitefly infestation (also caused by aphid). Besides causing lint stickiness, honeydew encourages the growth of moulds on the lint which then becomes black and valueless for processing.

Pectinophora gossypiella *(Saund.) (Lepidoptera: Gelichiidae).* Pink bollworm was first recorded on cotton in 1937 in Bandar Abbas in southern Iran and is thought to have originated in neighbouring countries. It has been found on wild cotton in Baluchestan. Losses in yield of up to 40 percent have been recorded. Because of its potential importance, internal and external quarantine measures against this pest are strictly enforced. Imports of cotton lint, seed cotton and cottonseed is prohibited.

Aphis gossypii *(Glov.) (Hemiptera: Aphididae).* Cotton aphid is found mainly on cotton, but also on other crops, and in Iran is known as the cucurbit aphid. It is common on horticultural crops. Damage to cotton is caused by direct feeding on the sap which weakens the plant and by the production of honeydew which reduces the fibre quality. Cotton aphid is probably the vector of some of the virus diseases found on vegetables.

Oxycarenus hyalinipennis *(Costa)* *(Hemiptera: Lygaeidae).* Cottonseed bug feeds on the seeds of cotton and other malvaceous hosts, but does not harm the cotton lint. Seed damaged by cottonseed bug has a reduced viability and consequently a loss of stand of up to 15 percent may occur. Between cotton seasons cottonseed bug occurs on malvaceous weeds and its numbers begin to increase in early spring. The pest moves on to cotton when the crop is well developed and females begin to oviposit on splitting bolls. Early harvest reduces the importance of cottonseed bug.

Nezara viridula *(L.)* *(Heteroptera: Pentatomidae).* Green grass or green stink bug attacks cotton, vegetable crops and tobacco in Iran. It can reduce yield by up to 45 percent as a result of its feeding. Lint from damaged bolls is immature and stained yellow-brown. It is a minor pest, regulated by natural enemies, especially the egg parasitoid *Trissolcus (syn. Micropha-nurus) basalis*(Wollaston) (Hymenoptera: Scelionidae), and chemical control is seldom necessary.

Diseases

Verticillium wilt. Verticillium wilt, caused by *Verticillium dahliae* Klebahn was first observed in the northern and northwestern cotton-growing areas in 1953. Damage varies with variety, climatic conditions and soil type. Cotton research institutes regularly screen new breeding material for tolerance or resistance to this disease. Coker 100 Wilt has been found to be susceptible; in 1985 verticillium wilt caused losses of up to 80 percent in this variety. Sahel, however, shows a good level of tolerance and is recommended for cultivation in those areas most infested with the disease, especially Mazandaran and Gorgan. The following practices are recommended to minimize the impact of the disease:
- crop rotation;
- weed control;
- irrigation practices;
- good drainage;
- healthy seed;
- use of fungicidal seed dressings;

- avoidance of mechanical damage to cotton during weeding;
- destruction of crop residues;
- avoidance of overuse of fertilizer;
- selection of resistant varieties.

Seedling diseases. Seedling damping-off, caused by *Rhizoctonia solani* Kuhn, occurs in the cooler, wetter regions, notably Gorgan, Moghan and Mazandaran. In favourable conditions up to 70 percent of seedlings may be affected. Sowing too early or too deep and temperatures in the range of 15 to 18°C encourage infection. At temperatures above 21°C the damage caused by damping-off diminishes rapidly.

Fusarium wilt. Fusarium wilt, caused by *Fusarium oxysporum* Schlecht f. sp. *vasinfectum* Atk. Sny. & Hans, is very serious on the Caspian Sea coast and in Esfahan and Khorasan. Up to 20 percent of plants may be lost and boll size and fibre quality reduced because of attack. The presence of root knot nematodes, *Meloidogyne* spp., increases the incidence of fusarium symptoms.

Weeds

The major weed species competing with cotton in Iran include: *Chenopodium album* L. (Chenopodiaceae), *Amaranthus* spp. (Amaranthaceae), *Solanum nigrum* L. (Solanaceae), *Malva parviflora* L. (Malvaceae), *Malva montana* Fork. (Malvaceae), *Hibiscus trionum* L. (Malvaceae), *Abutilon avicennae* Gaertn. (Malvaceae), *Echinochloa crus-galli* (L.) P. Beauv. (Gramineae) and *Setaria viridis* (L.) P. Beauv. (Gramineae).

CONTROL MEASURES
Chemical control

Chemical control is the principal method of pest control in Iran. Table 4 shows the area of cotton, in hectares, sprayed in the 1992 and 1993 seasons in the different cotton-growing regions, by aerial and ground methods. The chemical control strategy to be followed each season is decided by a committee comprising crop protection research scientists, other specialists

TABLE 4

Area sprayed for control of cotton pests in Iran in 1992 and 1993

Region	1992		1993	
	Aerial *(ha)*	Ground *(ha)*	Aerial *(ha)*	Ground *(ha)*
Gorgan	225 835	39 618	113 626	13 600
Khorasan	42 817	-	31 065	-
Mazandaran	56 584	2 005	39 371	65
Azarbaijan, Est.	35 142	2 086	7 000	-
Tehran	12 965	-	9 466	-
Fars	7 512	637	4 835	327
Semnan	5 465	-	3 028	-
Central	1 066	-	800	-
Subtotal	**387 386**	**44 346**	**209 191**	**13 992**
Total	**431 732**		**223 183**	

and crop protection professionals, who consider the latest results from pesticide evaluation trials. The committee selects the pesticides, including insecticides, herbicides and fungicides, to be used and the method of application and dosage rates. Table 5 shows the range of pesticides currently used on cotton, together with target pests, diseases and weeds and rates, method and number of applications. Spraying decisions are based on the results of regular pest monitoring carried out by crop protection specialists who visit the fields regularly. Severe outbreaks of a major pest may need to be dealt with in a manner different from normal, routine pest control operations.

Insecticide resistance. Levels of resistance to insecticides in major pest populations are continuously monitored and new products evaluated for their performance against resistant species. Ineffective control of a particular pest by a pesticide may have more than one cause. In whitefly, for example, ineffectiveness may be caused by resistance or by careless or inadequate application methods, including underdosing.

TABLE 5

Principal pests, diseases and weeds in cotton in Iran and recommended pesticides

	Pesticide and formulation	Rate per ha formulation	Number of applications	Notes
Pest				
Earias insulana	Carbaryl 85% wp	3.0 kg	1-4	
	Azinphos methyl 20% ec	5.0 litres	1-4	
Helicoverpa armigera	Thidiocarb 80% wg	1.0 kg	1	
	Profenofos 40% ec	2.0 litres	1	
	Carbaryl 85% wp	3.0 kg	1-2	
	Endosulphan 35% ec	3.0 litres	1	
Spodoptera exigua	Carbaryl 85% wp	30 kg	1-4	
Autographa gamma	Endosulphan 35% ec	3.0 litres	as necessary	
Spodoptera litura	Monocrotophos 40% sl	2.0 litres		
	Etrimphos 50% ec	1.5 - 2.0 litres		
Agrotis segetum	Lindane 25% wp	50 kg bait		
	Carbaryl 85% wp	50 kg bait		
Tetranychus urticae	Dicofol 18.5% ec	2.5 litres	1-2	Use of dicofol
	Propargite 57% ec	1.5 litres	1-2	restricted to crops
	Monocrotophos 40% sl	1.5 litres-2.01 litres	1-2	that are not directly
	Tetradifon 7.52% ec	4.0 litres	1-2	consumed
	Binapacryl 40% ec	1.5 litres	1-2	
	Binapacryl 50% wp	1.5 kg	1-2	
Thrips tabaci	Oxydemeton-methyl 25% ec	0.5-1.0 litres		
Aphis gossypii	Thiometon 25% ec	0.5-1.0 litres		
Empoasca sp.	Phosphamidon 50% sl	0.5-1.0 litres		
Oxycarenus sp.	Dimethoate 40% ec	0.5-1.0 litres		
Nezara viridula				
Bemisia tabaci	Carbaryl + 85% wp	3.0 kg	1-2	
	Phosphamidon 50% sp	1.0 kg	1-2	
	Endosulphan 35% ec	2.5-3.0 litres	1-2	
	Pirimiphos-methyl 50% ec	1.5 litres	1-2	
Diseases				
Rhizoctonia solani	Quintozene 75% wp	4-6 g per kg seed	1	Seed dressing
	Carboxin + thiram 75% wp		1	
Weeds				
Chenopodium album	Ethalfluralin 35.5% ec	3-4 litres	1	Pre-emergence
Amaranthus spp.	Trifluralin 48% ec	2.5-3 litres	1	herbicides
Solanum nigrum	Dinitramine 25%	3 litres	1	incorporated
Malva parviflora				into the soil
M. montana				
Hibiscus trionum				
Abutilon avicennae				
Echinochloa crus-galli				
Setaria viridis				

Integrated control. Investigations have been carried out to determine whether integrated control, especially varietal resistance, can replace the current reliance on pesticides for the control of sucking pests especially whitefly, jassid, spider mite and thrips. The varieties tested included Varamin 61, Sahel, GL4, Super Okra and Deltapine SL. Super Okra was found to be the most resistant to thrips, spider mite and jassid. All the varieties were equally susceptible to whitefly.

Environmental impact of chemical control. The use of pesticides on cotton has been observed to result in changes in the agro-ecosystem, including the resurgence of pest populations, the emergence of new pest species and the spread of pest populations from one major host plant to another. Whitefly exhibits all these features.

The ecological and social conditions in Gorgan and Gonbad make these regions particularly suitable for cotton production. Considerable quantities of pesticides are used annually in these regions with inevitable effects on the environment, especially the contamination of non-target areas, including water, by pesticides. The use of persistent organochlorine pesticides, which are particularly damaging to the environment by leaving residues in the soil, is now banned, except for lindane which is still used in baits.

Pest forecasting and monitoring networks. Decisions on the need to spray cotton, the type of pesticides to use and the frequency of application are based on regular monitoring of pest populations in cotton fields by a network of field scouts. As a result, farmers make judicious use of pesticides and it has been possible to reduce the number of sprays from between eight and ten per season to an average of three. In aerially sprayed areas monitoring teams observe weather conditions as well as pests to ensure that applications are made only in favourable weather. The application of insecticides only when pest populations reach economic threshold levels results in the economic and effective use of chemicals to protect crops and reduces the adverse effects of these chemicals on the agro-ecosystem and the environment generally.

Pesticide application. All ground spraying is carried out by farmers themselves and cotton areas of less than 10 ha are always ground sprayed. Larger areas are sprayed from the air, by the Special Services Aviation Company, owned by the Ministry of Agriculture. Farmers pay the cost of the spraying but the costs of pest monitoring and forecasting and of extension advice are borne by the government.

Government role. The state supports agriculture and ensures the supply of the inputs necessary for cotton production as well as providing information and advice to farmers. Specialists, professionals and technicians in the plant protection services play an essential part in the education and training of farmers and in directing control operations.

Under normal conditions the government is not directly involved in control operations, which are the responsibility of the farmer. Exceptionally, when there are large-scale pest outbreaks over a wide area, the government may declare an emergency. Farmers are then obliged to carry out specified control operations according to government instructions and recommendations. If farmers do not comply, the plant protection department carries out the control and recovers the cost from the farmer, through the courts if necessary. An outbreak of spiny bollworm in the coastal region of the Caspian Sea in 1966/67 was an example where this type of state intervention was necessary.

Private sector. Private-sector companies are also involved in cotton pest control and operate under the supervision and according to the regulations of the plant protection services. Foreign companies are also involved in aerial application of pesticides.

Legislative control

The campaign to control spiny bollworm in 1966-67 was carried out under the provisions of the Plant Protection Act (Articles VI and VII) whereby the Ministry of Agriculture is empowered to conduct nationwide control measures to prevent the spread of plant and storage pests and diseases and to require all rural cooperatives, farmers, owners and tenants to carry out control operations as directed by the Plant Protection Organization (OPP). OPP is required to inform the public of the type of pesticides to be used and the timing and place of their application, to prevent the poisoning of humans and livestock. Farmers unable to carry out the required control measures themselves may delegate this to OPP or the pest control companies, who will charge for the service.

Plant quarantine regulations include measures aimed at ensuring pink bollworm is not imported into the country.

Cultural control

The main practices aimed at reducing the impact of pests include:
- Deep ploughing after harvest is an important practice in the control of American bollworm. Pupae are either exposed on the surface to weather and predators or fail to emerge as adults. This practice can destroy up to 50 percent of pupae in the soil. Flooding the fields can destroy nearly all the pupae. Short-term pupae in the summer may also be destroyed by irrigation which compacts the soil and leads to their suffocation.
- Control of sowing dates – these have been set for each cotton-growing region by the Plant Breeding Institute of the Ministry of Agriculture and farmers are required to observe them.
- Field sanitation – farmers are required to destroy crop residues by burying them after harvest to reduce overwintering pest populations. Soil fertility is also increased by this practice.
- The movement of harvested cotton is controlled, as is that of ginnery waste, to ensure that pests are not transported from one area to another.
- Weeds and alternate host plants of pests and diseases are destroyed where possible.

Cultural control is important in suppressing whitefly; among the measures directed specifically at this pest are:
- avoidance of late-season irrigation in cotton;
- early harvest of cotton (this also reduces the impact of cottonseed bug) and, where possible, of cucurbits;
- destruction of cucurbit residues immediately after harvest;
- avoidance of the cultivation of cotton and cucurbits in adjacent fields;

- avoidance of the cultivation of sunflower or sesame on the edges of or within cotton fields;
- destruction of weeds that provide hosts during the winter.

Biological control
Research has shown that cotton pests have a number of natural enemies.

Predators. *Chrysoperla* spp. (Neuroptera: Chrysopidae) is widespread in most cotton-growing areas. It is predacious on bollworm eggs and larvae, including American bollworm (instars two to four) and spiny bollworm and usually produces four generations per year.

Cataglyphis sp. (Hymenoptera: Formicidae) is an early- and late-season predator of spiny bollworm larvae in the hot regions of southwestern Iran.

Parasitoids. *Microbracon* sp. (Hymenoptera: Braconidae) is a parasitoid of bollworm larvae. Oviposition may be directly on to the body of the host or eggs may be laid on bolls by the bollworm entrance hole. Development takes from 15 to 20 days and the adult may live for up to eight weeks.

Goryphus (syn. Brachycoryphus) nursei (Cameron) (Hymenoptera: Ichneumonidae) is a pupal parasitoid of spiny bollworm and other Lepidoptera.

Encarsia formosa Gahan (Hymenoptera: Aphelinidae) is a parasitoid of whitefly, especially greenhouse whitefly.

Encarsia inaron (Walker) (Hymenoptera: Aphelinidae) is a parasitoid of whitefly in the Tehran and Gorgan areas.

Eretmocerus mundus (Mercet) (Hymenoptera: Aphelinidae) is a parasitoid of whitefly that is important in the Tehran and Gorgan regions where up to 70 percent of whitefly pupae may be parasitized.

Trissolcus basalis (Wollaston) (Hymenoptera: Scelionidae) is an egg parasitoid of the green cotton bug *Nezara*.

Indirect control
Microbials. Microbial control agents tested experimentally include nuclear polyhedrosis virus (NPV) and the fungal pathogens *Metarhizium anisopliae*

(Metch.) Sorokin and *Beauvaria bassiana* (Bals.) Vuillemin for control of American bollworm.

In glasshouse conditions the fungus *Trichoderma harzianum* Rifai gave good control of *Rhizoctonia solani* on cotton, rice, soybean and sugar beet. *Pseudomonas fluorescens* (Trevisan) Migula which produces an antibiotic, pyrolitrin, gave significant control of *Rhizoctonia* and other fungi responsible for damping-off, including *Alternaria* and *Verticillium dahliae.*

Defoliation. Towards the end of the season, when whitefly-produced honeydew contaminates opening bolls and causes lint stickiness, the dense growth of cotton makes chemical control of whitefly ineffective. The use of defoliants, applied at 50 percent boll split, to desiccate and remove leaves infested with whitefly, protects the lint, facilitates harvesting and reduces the number of picks allowing earlier postharvest soil cultivation. The defoliants used are thidiazuron ("Dropp") or S,S,S-tributyl phosphoro-trithioate ("DEF 6"), either alone or mixed together.

INFRASTRUCTURAL SUPPORT FOR COTTON IPM RESEARCH
Research
Crop Improvement Research Institute. Activities include cotton breeding to produce new high-yielding, high-quality varieties appropriate to the conditions found in the different cotton-growing areas of Iran. The institute plays a decisive role in the development of varieties tolerant or resistant to the main cotton pests and diseases.

Plant Protection Research Institute. Research projects include studies on the biology, ecology and pest status of cotton pests and diseases and investigations on various methods of control including chemical control and IPM. Importance is given to the environmental impact of candidate insecticides, fungicides and herbicides being tested, as well as to their effectiveness against target organisms. Long-term investigations are in progress on biological control and control by natural enemies.

Operations

The role of the Plant Protection Organization (OPP) is defined in the relevant legislation covering plant protection and is executed through the network for pest forecasting, surveillance and control, which covers the different agricultural areas. The headquarters of OPP are in Tehran, with branches in each province and district, down to the level of small villages. The organization, which includes specialists, professionals, technical assistants and administrators on its staff, is thus widely distributed throughout Iran.

In the protection of crops and orchards OPP has two principal responsibilities; plant quarantine and the control of general pests, especially rodents, locusts and sunn pest, *Eurygaster integriceps* Puton (Hemiptera: Scutelleridae), on cereals. The government pays the cost of controlling these pests. A third responsibility concerns the maintenance of the pest monitoring and forecasting network and the dissemination of information and advice on control methods for the main crops and crops of strategic importance.

The role and activities of OPP, especially the control of *Eurygaster,* which is the main pest of wheat and barley, are very important for the national economy. Another important role of OPP relates to pesticide regulation. All decisions concerning pesticides, including selection of product, formulation, dosage and method of application for a particular purpose, are the responsibility of OPP.

Extension services

The Cotton and Oilseeds Organization. This organization is concerned with the production of cotton and oilseeds and has its central administration at Gorgan with branches in all the cotton-growing districts. It arranges contracts with growers and provides the services necessary for production of the crop. Its links with the research institutes of the Ministry of Agriculture allow it to disseminate information on new varieties and production techniques in the different cotton-growing areas. It covers all aspects of cotton production, including the purchase of seed cotton, ginning and distribution to end-users in the textile and other industries.

The agricultural extension service. The general extension service is concerned, at the farmer level, with the production of all crops. It provides advice on land preparation and demonstrations of different cultural methods and of the use of various types of agricultural machinery. The headquarters are in Tehran with offices in all districts.

Private sector. Apart from an aerial spraying company, which is responsible to the Ministry of Agriculture, there is no private-sector involvement in pest control operations. There are private-sector pesticide dealers, however, who trade under the authority and control of OPP.

FOREIGN ASSISTANCE TO COTTON PEST MANAGEMENT

The five-year plan for the economic development of Iran included a national project for cotton development which started in 1989 and ran for five years. The project had several components and coordinated the various organizations concerned to encourage the development of cotton production. The project included a subproject to develop and expand the forecasting and monitoring network for cotton pests and diseases.

The Cotton and Oilseeds Organization maintains scientific and technical links with overseas cotton research scientists and institutes and cotton producers, as well as with international organizations concerned with cotton. Information, materials and experts are exchanged.

KEY PESTS

The whitefly *Bemisia tabaci* is the key pest of cotton in Iran.

KEY PERSONNEL INVOLVED IN COTTON PEST MANAGEMENT IN IRAN

Dr H. Bayat-Assadi is an entomologist, currently Head of the Division of Biological Control at the Research Institute for Plant Pests in Tehran. For many years he worked on cotton pests in the Gorgan region, one of the most important cotton-growing areas of Iran.

M.Ing. Glitchabai is a scientist working on cotton pests at the laboratory for research on plant pests at Gorgan.

M.Ing. Kaviani is responsible for crop protection in Gorgan.

M.Ing. A.A. Mousef is an entomologist with 20 years experience of cotton pest research, mainly in the south of the country at Fars. He is currently based in Shiraz.

There are many other people involved who have long-term field experience of cotton pest management, including pest monitoring and forecasting and control operations.

References

Bamedadian, A. 1987. *Lutte biologique avec certains agents (champignons).* L'Institut de recherches des ravageurs et des maladies des plantes. Tehran, Iran, OPP.

Behbahani, M., & Beyat-Assadi, H. 1987. L'application des insecticides microbiens (virose, champignons, et protozoaires) vis-à-vis des différents ravageurs. 1ére Seminaire de la Lutte Biologique Contre des Ravageurs et des Maladies des Plantes, Tehran, Iran, IRRMP/OPP.

Behdad, E. 1982. *Pests of field crops in Iran.* Tehran, Iran, Sepehr Ed.

Djavan-Mogadam, H. & Bachar, G. 1991. Etude de la contamination des variétés de coton aux insectes suceurs. 10ème Congrès de la Protection des Plantes d'Iran, 1 to 5 September 1991, Kerman, Iran.

Madjidieh-Gassemi, Ch. & Hassanzadeh, N. 1987. Contrôle de *Rhizoctonia solani* sur le coton avec *Pseudomonas fluorescens.* 1ère Seminaire de la lutte biologique contre des ravageurs et des maladies des plantes, Tehran, Iran, IRRMP/OPP.

Moshir, Abadi. 1983. *Verticillium dahliae* in Iran. *Proceedings of the 7th Plant Protection Congress of Iran.* College of Agriculture, University of Tehran, Karadj, Iran.

Iraq

H.S. El-Haidari

INTRODUCTION

Commercial cotton growing began in Iraq in 1920 (Walker, 1953) following several years of variety trials conducted by the Department of Agriculture. From these experiments it was concluded that American upland varieties *(Gossypium hirsutum)* would be the most suitable for Iraq. In the late 1940s high yields and favourable prices encouraged cotton production and the area under the crop reached a peak of 113 000 ha in 1951. That year, however, prices and yields declined – the latter because of severe spiny bollworm damage – and in 1952 the cotton area contracted to about 50 000 ha.

Production continued to decline and by the 1980s averaged only some 13 000 tonnes of seed cotton per year; Table 6 gives data for the period 1981 to 1989. One reason for this decline was a shift from cotton to short-season summer cash crops such as maize, potatoes, sunflower and rice, which give farmers good financial returns within a short period compared with cotton, which is a long-season crop with high labour demands. Government efforts to counter the decline in cotton production through the provision of various incentives have not met with success. The most important areas for cotton production are now located in Ta'meen, Diala and Salah Al-Deen provinces (Government of Iraq; Annual Abstract of Statistics, 1989).

CULTURAL PRACTICES

In the central zone land preparation for cotton begins from early to mid-March; in the northern areas from mid- to late March. Ploughing is by mould-board to a depth of 25 cm and this is followed by disc harrowing to 15 cm depth. Land planes are used for levelling. In the central zone subsoiling with a chisel plough to a depth of 50 cm is recommended every

TABLE 6
Area, production and yield of seed cotton in Iraq, 1981 to 1990

Year	Area (ha)	Production (tonnes)	Yield (kg per ha)
1981	11 350	13 300	1 168
1982	12 050	14 100	1 172
1983	13 750	11 800	860
1984	9 975	7 100	716
1985	10 825	7 200	668
1986	2 3050	20 300	880
1987	18 650	14 200	763
1988	13 275	12 200	920
1989	7 400	14 400	1 947

four years. Recommended fertilizer applications comprise 70 kg P_2O_5 per hectare broadcast before harrowing and 90 kg N per hectare split between two applications. If machinery is used fertilizers can be applied in bands, 5 cm to the side of the row and 5 cm below the surface. A total of 15 to 16 irrigations are required, the first before sowing, with subsequent irrigations every 15 days until mid-June, after which the interval is reduced to ten days until the end of August. In September one or two further irrigations are recommended to finish the crop. On small farms cotton is hand-picked; on larger farms machine harvesters are used and harvesting starts ten days after defoliation with "DEF 6" (S,S,S-tributyl phosphorotrithioate), either alone or in combination with paraquat. Following harvest, incorporation of cotton stubble into the soil is recommended to reduce pest carryover to the next season.

The main commercial variety now grown in Iraq is Coker 310 which replaced Coker 100 Wilt because of its tolerance to verticillium wilt and its improved ginning percentage.

PESTS
Insects
About 24 species of phytophagous arthropods have been reported on cotton in Iraq (Table 7), the most important being spiny bollworm and strawberry spider mite followed by whitefly. Brief notes on the main pest species follow.

Earias insulana (Boisd.) *(Lepidoptera: Noctuidae).* Spiny bollworm was first reported in Iraq by Rao (1921). Walker (1953) estimated yield losses of about 80 percent to this pest at one site (Abu-Gharaib farm) and it has become one of the major causes of the decline cotton production. Insecticides, applied by the farmer using ground application equipment or by helicopter hired from the Ministry of Agriculture, are the main method of control. Currently the following insecticides are recommended by the ministry against spiny bollworm: azinphos-methyl 30% ec at 3.0 litres per hectare; chlorpyrifos 24% ul at 4.0 litres per hectare; fenvalerate 7.5% ul at 2.0 litres per hectare; methidathion 40% ec at 2.4 litres per hectare; monocrotophos 60% at 2.0 litres per hectare; and phenthoate 92% ul at 2.0 litres per hectare.

Treatment begins at the first appearance of infestation and is repeated every 15 days with fairly good results, although it is now suspected that spiny bollworm is resistant to some insecticides. Environmental pollution by pesticides is becoming a problem.

Non-chemical methods of control have not been developed and, although the destruction of crop residues is required by law, few farmers abide by this. The practice of using uniform sowing dates over a wide area concentrates the fruiting period and limits the development of populations of spiny bollworm and other pests. Pre-sowing irrigation has been shown to increase plant populations by 71 percent and seed cotton yields by 41 percent. Walker (1953) observed parasitism by *Bracon hebetor* Say (Hymenoptera: Braconidae) but its importance in control is not known. There has been no attempt to assess different cotton varieties for resistance to insect or mite pests in Iraq.

TABLE 7

Insects and mites recorded on cotton in Iraq

Species		Status
(Lepidoptera: Noctuidae)	*Earias insulana*(Boisd.) *Helicoverpa armigera* (Hb.) *Spodoptera littoralis*(Boisd.) *S. exigua* (Hb.) *Xanthodes graellsii*(Feisth.)	+++ ? ? ? ?
(Lepidoptera: Gelechiidae)	*Pectinophora gossypiella* (Saund.)	?
(Lepidoptera: Arctiidae)	*Utetheisa pulchella* (L.)	?
(Lepidoptera: Cosmopterigidae)	*Sathrobota simplex* (syn. *Pyroderces*) (Wals.)	?
(Thysanoptera: Thripidae)	*Thrips tabaci* (Lind.) *Frankliniella pallida*(Uzel) *F. intonsa* (Trybom)	+ ? ?
(Hemiptera: Aleyrodidae)	*Bemisia tabaci* (Genn.)	++
(Hemiptera: Aphididae)	*Aphis gossypii* (Glov.)	+
(Hemiptera: Cicadellidae)	*Empoasca decedens* (Paoli)	+
(Hemiptera: Pentatomidae)	*Nezara viridula* (L.) *Acrosternum* sp. *Eurydema ornata* (L.)	+ + ?
(Hemiptera: Miridae)	*Deraeocoris ostentans* (Stal.) *Phytocoris* sp.	? ?
(Hemiptera: Lygaeidae)	*Oxycarenus hyalinipennis* (Costa) *Spilostethus pandurus*(Scopoli)	+ ?
(Coleoptera: Scarabaeidae)	*Oxythrea cinctella* (Schaum)	?
(Isoptera: Termitidae)	*Microcerotermes diversus* (Silvestri)	
(Acari: Tetranychidae) (Ugarov and Nikol'skii)	*Tetranychus turkestani*	+++

+++ very important;
++ moderately important;
 + of minor importance;
? present but of unknown importance.

Helicoverpa armigera *(Hb.) (Lepidoptera: Noctuidae).* Walker (1953) reported collecting one larva of American bollworm from cotton at Abu-Gharaib farm in September 1952. Some years later, in 1962, at the state cotton farm some 70 km south of Baghdad, the author noted a serious outbreak which resulted in a heavy loss of yield. Another outbreak occurred at Abu-Gharaib farm in 1964. Since then, however, American bollworm has not been of any economic importance.

Pectinophora gossypiella *(Saund.) (Lepidoptera: Gelechiidae).* Pink bollworm was first reported from Al-Anbar province, near the Syrian border, in 1968 (El-Haidari, Naoum and Abid, 1969). It was subsequently observed in some areas of Baghdad, Wasit and Babil provinces but its spread to other provinces appears to have been successfully prevented by plant quarantine measures and it is now of no economic importance.

Other Lepidoptera. *Spodoptera littoralis* (Boisd.), *S. exigua* (Hb.), *Xanthodes graellsii* (Feisthamel) (all Lepidoptera: Noctuidae) and *Utetheisa pulchella* (L.) (Lepidoptera; Arctiidae) have all been recorded in Iraq but at present are of no economic importance on cotton.

Bemisia tabaci *(Genn.) (Hemiptera: Aleyrodidae).* The cotton whitefly has been recorded on 61 plant species in Iraq (Al-Janabi, 1986). A light infestation was observed on cotton at Abu-Gharaib farm in 1952 (Walker, 1953). At present it is considered to be of moderate importance. The application of insecticides to control spiny bollworm probably has adverse effects on *Eretmocerus* spp. and *Encarsia* spp. (syn. *Prospaltella*) (Aphelinidae), parasitoids of whitefly which exercise some control. Whitefly infestations on cotton are first observed in June and build up to peaks in August or September (El-Haidari *et al.,* 1983). Where chemical control is needed, malathion 50% ec at 2 litres per hectare is recommended. Whitefly infestation is encouraged by overuse of fertilizer and irrigation water and by very high plant populations. The elimination of alternative weed hosts is recommended.

Aphis gossypii *Glov. (Hemiptera: Aphididae).* Aphid infestations start on cotton at the seedling stage and populations peak in the first week of May, usually declining thereafter as temperatures rise although populations may persist throughout the season. Natural enemies, including the predator *Coccinella septempunctata*Linnaeus (Coleoptera: Coccinellidae), exercise some control. Insecticides are not used against this pest.

Empoasca decedens*Paoli (Hemiptera: Cicadellidae).* Nymphs and adults of the cotton jassid have been reported feeding on cotton during May, often continuing throughout the season (El-Haidari, 1964). The damage caused, however, is of no economic importance and insecticides are not used against this pest.

Other Hemiptera. A number of Pentatomid bugs, including*Nezara viridula* (L.), *Acrosternum* sp. and *Eurydema ornata* (L.) (Hemiptera: Pentato-midae), have been recorded feeding on cotton buds, flowers and bolls (El-Haidari, 1964) but are not considered to be of economic importance. Two Lygaeid bugs, *Oxycarenus hyalinipennis* (Costa) and *Spilostethus* (syn. *Lygaeus) pandurus* Scopoli, have also been recorded but are of no economic importance.

**Thrips tabaci *Lind.* (*Thysanoptera: Thripidae).* In Iraq thrips is a major pest of onions, but is of only minor importance on cotton where it may attack seedlings, especially where cotton is grown in vegetable-producing areas. Seedlings usually recover from early infestations (Al-Faisali, 1981). Two other thrips, *Frankliniella pallida* (Uzel) and *F. intonsa* (Trybom) have also been collected on cotton (El-Haidari, 1964; El Haidari and Daoud, 1971) but their importance is not known. No control measures are practised against thrips.

**Tetranychus turkestani *Ugarov and Nikol'skii (Acari: Tetranychidae).* Outbreaks of the strawberry spider mite on cotton following the use of organophosphorus insecticides to control spiny bollworm were reported by Walker (1953), Walker and El-Haidari (1954) and Al-Azawi (1966). Al-Neamy (1979) reported that these insecticides increased mite fecundity and longevity and reduced populations of*Scolothrips sexmaculatus* (Pergande) and the mite *Pronematus* sp. (Acari: Tydeidae), predators of *Tetranychus*. Under laboratory conditions*Orius albidipennis*(Reut.) (Anthocoridae) was found to be an effective predator (El-Haidari and Georgis, 1977).

The following acaricides are recommended for control of strawberry spider mite, spraying commencing when one moving stage is observed per leaf: dicofol 18.5% ec at 4.0 litres per hectare; Neotox 12.5% ec at 4.0 litres per hectare; sulphur dp at 16 to 20 kg per hectare; sulphur wp at 8.0 kg per hectare; and tetradiphon 8% ec at 4.0 litres per hectare.

Diseases

Verticillium wilt. Verticillium wilt, *Verticillium dahliae* Kleb., was first reported on cotton in Iraq in 1973 in Nineveh and Arbil provinces (Mamluk, 1974). Surveys conducted in 1976-77 showed that the disease was more or less confined to these two provinces, where it is now considered to be a major problem, although it also occurs to some extent in Al-Anbar province. Infection rates in Nineveh can reach over 82 percent with the highest average being 75 percent in Battit, northeast of Nineveh. The highest infection levels occurred in fields that had been in continuous cotton cultivation for some years or where cotton was sown very late. Ahmed (1979) screened 62 cotton cultivars under glasshouse conditions and found some resistance to verticillium in some cultivars. Al-Beldawi *et al.,* (1983) investigated the susceptibility of different cotton cultivars to verticillium under field conditions. Coker 310, the main commercial variety, showed tolerance to the disease. The verticillium problem is being contained through plant quarantine laws which prohibit the transport of cottonseed from infested areas to disease-free areas. Farmers are advised not to grow cotton in the same field for more than two or three seasons to avoid the build-up of the disease in the soil.

Rhizoctonia solani *Kuhn.* Seedling damping-off caused by *Rhizoctonia* is considered important in Iraq. Al-Beldawi and Waleed (1985) screened 33 cotton cultivars under glasshouse conditions using artificially infested soil and investigated control measures. The experiments showed that the disease could be controlled by using fungicidal seed dressings, such as carboxin, at the rate of 4 to 5 g per kilogram of seed.

Fusarium wilt. Fusarium wilt, caused by *Fusarium oxysporum* Schlecht f. sp. *vasinfectum* Atk. Sny. and Hans, was first recorded in Iraq by Hussain (1965). It is of minor importance and no control measures are needed.

Anthracnose. Anthracnose, caused by *Glomerella gossypii*(Southw.) Edg., is observed on cotton late in the season but is of no economic importance at present and no control measures are needed.

Boll rots. Several fungi cause various types of boll rot and lint contamination of which the principal species are *Rhizopus stolonifer* (Ehrenb. ex Fr.) Lind. (syn. *R. nigricans*)*, Aspergillus niger* Van Teigh. and *Nematospora coryli* Peglion (Walker, 1953). Insect attack, especially by spiny bollworm, increases the incidence of boll rots.

Nematodes. The fungal-feeding nematode*Aphelenchus avenae*Bastion has been found associated with cotton. A species of *Pratylenchus* has also been recorded, but appears to be of minor significance (Hussain, 1965).

Weeds. Weed competition is a major factor limiting cotton yields in Iraq, especially during the first four weeks after sowing (Al-Kaisi, 1972). Kadhim *et al.* (1983) reported yield losses to weeds ranging from 38 to 51 percent at a series of sites in different provinces. Table 8 shows the main weed species at Abu-Gharaib. Mechanical cultivation is the most common method of weed control. Trifluralin 44.5% ec, used as a pre-emergence herbicide incorporated by disc harrow or rotary cultivator, gives good control of a range of broad-leaved and annual grass weeds.

Pest introductions. Pink bollworm (El-Haidari*et al.,*1969) and verticillium wilt (Mamluk, 1974) are the two most recent additions to the cotton pest complex.

INFRASTRUCTURAL SUPPORT FOR COTTON IPM
The implementation of the results of cotton pest management research is currently severely constrained by shortages of trained crop protection personnel, transport, equipment, pesticides and other facilities.

Plant Protection Research Centre. The centre, which is part of the State Board for Applied Agricultural Research of the Ministry of Agriculture, is

TABLE 8
Weeds of cotton at Abu-Gharaib, Iraq

Species	Status
Echinochloa colona(L.) Link (Gramineae)	+++
*Polygonum aviculare*L. (Polygonaceae)	+++
*Euphorbia helioscopia*L. (Boraginaceae)	++
*Amaranthus retroflexus*L. (Amaranthaceae)	++
*Centaurea pallescens*Del. (Compositae)	+
Sorghum halepense (L.) Pers. (Gramineae)	+
*Solanum nigrum*L. (Solanaceae)	+
*Xanthium strumarium*L. (Compositae)	+
Prosopis farcta (Banks and Sol.) Macbride (Leguminosae)	+
Alhagi maurorum(Leguminosae)	+
*Portulaca oleracea*L. (Portulacaceae)	+
Ammi majus L. (Umbelliferae)	+

+++ very common;
++ common;
+ infrequent.

responsible for research on arthropod pests, diseases and weeds on various crops, including cotton.

Department of Plant Protection. The department is part of the Agricultural Services Division of the Ministry of Agriculture. In cooperation with the State Board for Applied Agricultural Research and the state company for agricultural supplies, it is responsible for estimating pesticide requirements for cotton.

Department of Extension. Also part of the Ministry of Agriculture's Agricultural Services Division, this department has a very limited role in cotton pest management.

Private agencies. There are no private agencies involved at present.

Government programmes. Government organizations are responsible for cotton pest surveys, monitoring and forecasting, and for research on pest biology, with particular emphasis on control. The results of research are published annually by the Ministry of Agriculture in *The Yearbook of Plant Protection Research.*

Foreign assistance. There are no donor-assisted programmes at present.

KEY PESTS
Iraq is comparatively free of major pests of cotton. The main pests are: spiny bollworm (*Earias insulana*), strawberry spider mite (*Tetranychus turkestani*), whitefly (*Bemisia tabaci*), verticillium wilt (*Verticillium dahliae*), damping-off (*Rhizoctonia solani*) and weeds.

RECOMMENDATIONS
The following recommendations should be kept in mind by those involved in cotton pest control in Iraq:
- The IPM programme requires a wide range of skilled personnel, including research workers and extension and training specialists, if it is to be adopted by farmers.
- Further reduction in pesticide use will depend on research to determine economic threshold levels for the various pest species.
- The use of pheromones for spiny bollworm control and pest monitoring and population studies needs to be assessed.
- Research is required on the natural enemies of cotton pests to determine their role in control. Exotic natural enemies from elsewhere should be imported and evaluated as control agents, under quarantine conditions.
- External and internal quarantine regulations need to be enforced to ensure that new pest species are not introduced into the country and that pests already present in some areas are not spread to non-infested areas.
- Research on cotton agronomy and cotton breeding should be integrated with crop protection research to ensure a multidisciplinary approach to developing IPM systems, in particular to determine factors, including soil fertility and variety, which may affect pest numbers and behaviour.

• The effect of plant growth stage on pest populations in relation to control and to crop loss requires further research.

KEY PERSONNEL INVOLVED IN COTTON PEST MANAGEMENT IN IRAQ

Dr A.S. Al-Beldawi, Plant Pathologist, Plant Protection Research Centre, Abu-Gharaib.

Dr Faud K. Ismail, Herbicides Specialist, Plant Protection Research Centre, Abu-Gharaib.

Dr Ghaib A.W. Wael, Head Entomologist, Plant Protection Research Centre, Abu-Gharaib.

References

Ahmed, D. & Anis, N.A. 1960. *Chemical control experiments on the spiny bollworm* Earias insulana *Boisd. and the spider mite* Tetranychus *sp. on cotton in Iraq.* Technical Bulletin No. I, Division of Entomology, Ministry of Agriculture, Iraq. (In Arabic)

Ahmed, K.K. 1979. Screening cotton cultivars against verticillium wilt. University of Mosul, Mosul, Iraq. (M.Sc. thesis)

Al-Azawi, A.F. 1966. The effect of several insecticides on the red spider mite of cotton *Tetranychus atlanticus* McG. in Abu-Ghraib, Iraq. 5th Arab Science Conference, Baghdad, Iraq.

Al-Beldawi, A.S. & Waleed, B.K. 1985. Susceptibility of some cotton cultivars to *Rhizoctonia solani* Kuhn. *The Yearbook of Plant Protection Research,* 3(2): 255-261. Ministry of Agriculture, Baghdad, Iraq. (In Arabic)

Al-Beldawi, A.S. *et al.* 1983. Screening different cotton cultivars against verticillium wilt. *Yearbook of Plant Protection Research,* 3(2): 241-253. Ministry of Agriculture, Baghdad, Iraq. (In Arabic)

Al-Faisali, A.H.M. 1981. Ecological studies on onion thrips *Thrips tabaci* Lind. College of Science, University of Baghdad, Baghdad, Iraq. (M.Sc. thesis) (In Arabic)

Al-Janabi, S.D. 1986. Biology of the whitefly *Bemisia tabaci* (Genn.) in the middle of Iraq. College of Science, University of Baghdad, Baghdad, Iraq. (M.Sc. thesis) (In Arabic)

Al-Kaisi, K.M. 1972. *Weeds and their control.* Circular No. 62, Director-General of Field Crops, Ministry of Agriculture, Baghdad, Iraq. (In Arabic)

Al-Neamy, K.T. 1979. The positive effect of insecticides on the biology of strawberry spider mite *Tetranychus turkestani* Ugarov & Nikol'skii. College of Agriculture, University of Baghdad, Baghdad, Iraq. (M.Sc. thesis) (In Arabic)

El-Haidari, H.S. 1964. *Cotton insects.* Technical Bulletin No. 19, Director-General of Research and Agricultural Projects, Ministry of Agriculture, Baghdad, Iraq. (In Arabic)

El-Haidari, H.S. & Daoud, A.A.K. 1971. On a collection of thrips from Iraq. *Nat. Hist. Mus.,* 5 (1): 23-25. Baghdad, Iraq.

El-Haidari, H.S. & Georgis, R. 1977. Predation of *Orius albidipennis* Reuter (Hemiptera, Anthocoridae) on *Tetranychus atlanticus* McG. (Acari, Tetranychidae) in Iraq. *J. Agnew. Entomol.,* 83(3): 257-260.

El-Haidari, H.S., Naoum, A.N. & Abid, M.K. 1969. *Outbreaks and new record: pink bollworm on cotton.* FAO Plant Protection Bulletin No. 17 (6), Rome, Italy, FAO.

El-Haidari, H.S. *et al.* 1983. Population densities of *Bemisia tabaci* (Genn.) and *Thrips tabaci* Lind. on cotton, cucumber and okra. *The Yearbook of Plant Protection Research,* 3(I): 3-129. Ministry of Agriculture, Baghdad, Iraq. (In Arabic)

Hussain, F.H. 1965. *List of common plant diseases in Iraq.* Technical Bulletin No. 3, Director-General of Research and Agricultural Projects, Ministry of Agriculture, Baghdad, Iraq.

Kadhim, F. *et al.* 1982. Weed control in cotton fields by selective herbicides. *Yearbook of Plant Protection Research,* 2(2):269-271. Ministry of Agriculture, Baghdad, Iraq. (In Arabic)

Kadhim, F. *et al.* 1983. The effects of some selective herbicides on weeds in cotton fields. *Yearbook of Plant Protection Research,* 3(2): 453-464. Ministry of Agriculture, Baghdad, Iraq. (In Arabic)

Mamluk, O.F. 1974. A first report on verticillium wilt of cotton in Iraq. *Plant Disease Reporter,* 58: 996-997.

Rao, Y.R. 1921. *A preliminary list of insect pests in Iraq.* Memoir No. 7, Agricultural Department, 35 pp.

Walker, R.L. 1953. *Report to the Government of Iraq on the control of the spiny bollworm.* FAO/53/7/5359. Rome, Italy, FAO.

Walker, R.L. & El-Haidari, H.S. 1954. Effectiveness of certain insecticides against the spiny bollworm in Iraq. *J. Econ. Entomol.,* 47(2): 367-369.

Morocco

L. El Jadd

INTRODUCTION

Cotton is the only fibre of vegetable origin that is cultivated to any extent in Morocco at the present time. It was first introduced into the country in 1914 when the National Institute for Agronomic Research (INRA) commenced variety trials at Rabat. In 1951 a cotton research station was established at Afourer in collaboration with the Compagnie Français pour le Developpement des Fibres Textiles (CFDT) and the Institut National du Coton et Textiles Exotiques (IRCT).

All the cotton grown is long-staple *Gossypium barbadense* and the main varieties include Pima 67 an Tadla 16. Table 9 shows the general production data for the main production areas.

National production has varied from 3 000 to 11 000 tonnes of lint per year since the mid-1980s. The 1987/88 season being the peak year for this period (ICAC, 1993).

The local textile industry consumes more than 30 000 tonnes of cotton annually, much of this demand being met by imports of medium-staple

TABLE 9

Area, production and yield of cotton in Morocco for 1992-93

Region	Area ('000 ha)	Production ('000 tonnes)	Yield of lint (kg per ha)	Months	Cotton season (days)
Beni Mallal (Tadla)	2.1	2.5	410	May/Apr. to Sept./Oct.	270
Doukkala	1.1	-	-	Apr./Jun. to Sept./Dec.	210
Gharb	1.2	0.6	470	May to Oct./Nov.	180
Haouz	0.8	0.5	570	May to Oct./Nov.	180

Source: Survey of Cotton Production Practices, ICAC, 1993.

cotton, while locally produced high-value long-staple cotton is exported. Cottonseed is crushed to provide cooking oil and cottonseed cake for animal feed.

CULTURAL PRACTICES

In Gharb more than 90 percent of farms are less than 5 ha, and in Doukkala and Haouz 70 percent are less than 10 ha. In Beni-Mellal (Tadla) one-quarter of farms are over 20 ha, while of the remainder one-third are less than 5 ha and a further quarter range from 5 to 10 ha. Cotton is rotated with cereals, vegetables and lucerne, or may follow a fallow period. Soil cultivation is almost entirely mechanized; irrigation is on the ridge and furrow system. Harvesting is entirely by hand and the seed cotton is roller ginned (ICAC, 1993).

PESTS

Insect pests causing problems on cotton in Morocco can be divided into two categories; those pests, mainly bollworms and leafworms, that can cause loss of yield, and the sucking pests, notably whitefly and aphid, which secrete honeydew, thus causing loss of quality. In 1975 cotton pest problems in Morocco were reviewed and it was concluded that Lepidoptera were the main cause of yield loss for reasons that included:
- intensification of crop production on a limited area of irrigated land;
- increased diversity of cropping, leading to a succession of suitable host plants for Lepidopterous larvae;
- misuse of insecticides and reliance on calendar spraying, which led to insecticide resistance in certain pests.

As a result of this review an increased programme of research was started on cotton pests and their control, including trials on the efficacy of various insecticides, conducted on small-scale (12-ha) trial sites and large-scale (48-ha) sites. Studies on the biology and ecology of pests included the use of light and pheromone traps, sampling on various crops (both sprayed and unsprayed) and laboratory investigations on pest physiology.

Earias insulana *(Boisd.) (Lepidoptera: Noctuidae).* Spiny bollworm is a major mid-season pest of cotton in Morocco where it is widely distributed.

Early in the season (in May in Tadla), before fruiting points are present on the cotton, spiny bollworm bores into the growing points on the main stem of the cotton plant. The economic importance of this early season damage depends on the size of the population that survives the winter on alternate host plants. In mild winters spiny bollworm may go through two generations before moving on to cotton and tip-boring may be severe. Cold winters result in far smaller initial populations on cotton. The second generation develops on cotton flower buds in June and July. The date of sowing and early summer temperatures determine the size of this generation. The third generation is more concentrated in time and attacks bolls in August. A fourth generation may occur in September/October, before oviposition switches to alternate host plants.

Pectinophora gossypiella *(Saund.) (Lepidoptera: Gelechiidae).* Pink bollworm is an important pest of cotton in all areas, but especially in the Gharb region. It spends the winter as a diapause fourth-instar larva in cottonseed that is stored on or has fallen to the ground during harvest, or in bolls remaining on cotton stalks. Adults begin emerging from diapause in spring when there is no cotton at a suitable stage for oviposition and larval development. Consequently, only the later moths to emerge from diapause are able to breed successfully, once the cotton has flower buds and flowers. This first generation on cotton does little economic damage because of cotton's ability to compensate for the loss of early fruiting points. It is the second generation on cotton, which appears in July/August, that causes economic damage. There are two or three overlapping generations on cotton depending on the season.

Helicoverpa armigera *(Hb.) (Lepidoptera: Noctuidae).* American bollworm is a polyphagous pest that attacks a large range of cultivated plants throughout Morocco. On cotton it has been most important as a mid-season pest, especially in the Haouz region. There are usually three overlapping generations on cotton, the first from mid-June to mid-July, the second from mid-July to the beginning of August and the third from mid-August to the end of September. The first two generations cause most of the yield losses.

By destroying buds during the first half of the flowering period, this pest also delays the onset of boll splitting, often until after the start of autumn rains. Older plants are less attractive for oviposition, so if cotton is protected during the main period of flowering and boll loss is minimized, the bollworm moves on to other crops, notably tomatoes. Improvements in American bollworm control in recent years have led to this pest declining in importance and in some areas, such as Tadla, it is now almost non-existent.

Spodoptera littoralis *(Boisd.) (Lepidoptera: Noctuidae).* Surveys using pheromone traps have shown that the Egyptian cotton leafworm is distributed throughout the year in the Atlantic and Mediterranean coastal regions but disappears in winter in areas with a continental climate, such as Tadla. Of all the cotton Lepidoptera, it is the species most commonly caught in light traps. Plant sampling for late-stage larvae tends to underestimate the population because larvae drop off the plant when it is shaken.

   In areas of continental climate the origin of infestations on cotton is either hibernating larvae or immigration from other areas. Where the larvae are present throughout the year, winter survival is usually on lucerne. There are usually five to seven overlapping generations at Tadla, the first one is generally unnoticed and occurs on sugar beet. Infestations in July and August cause most damage to cotton, especially where the crop is grown in fields adjacent to lucerne.

Aphis gossypii *Glov. (Hemiptera: Aphididae).* Aphid is the most important early-season sucking pest in all the cotton-growing regions of Morocco. The optimum temperature range for the development of aphid populations is 15 to 26°C; rates of population increase are lower at temperatures outside this range and cease altogether at temperatures below 3°C. There are about 15 generations of this pest on cotton in the Tadla area. Late-season infestations can lead to a reduction in lint quality through stickiness and mould growth caused by honeydew exudates from the aphid.

Bemisia tabaci *Genn. (Hemiptera: Aleyrodidae).* Whitefly is becoming increasingly serious as a mid- to late-season pest, possibly as a result of increased use of insecticides against Lepidopterous pests. During the non-cotton season whitefly breeds on a large number of alternate host plants, especially weeds such as *Solanum* spp. and *Convolvulus* spp. There are several generations on cotton and loss of lint quality, caused by the effects of honeydew exudates, as with aphid, is the most significant damage.

CONTROL MEASURES
Chemical control

Decisions to use insecticides are based mainly on the use of economic thresholds in the main cotton-growing areas, although calendar-based spraying is still employed in Doukkala. Regional technical committees, with representatives from INRA, the Plant Protection Department (PV), farmers' associations and the Agricultural Products Marketing Company (COMAPRA), meet weekly during the season to decide which pesticides should be applied and at what rate of application.

New pesticides are evaluated in trials at research stations using plots of 72 m^2, replicated five or six times, followed by large-scale district trials covering 2 to 12 ha. Candidate pesticides are compared with the standard treatment against target pests and a range of application rates may be tested as well, the object being to select the most economic and effective control methods. Following these trials pesticides go through a final stage of validation in demonstration plots before being reviewed and, if appropriate, approved for use by the technical committees.

There has been a considerable decrease in the quantity of insecticide active ingredient used on cotton over the past two decades. Between 1973 and 1990 in Tadla region the average number of spraying applications declined from a peak of ten in 1977 to six in 1990. Organochlorine insecticides are no longer used and there has been a steep decline in the use of organophosphates and carbamates. To a certain extent these groups have been replaced by synthetic pyrethroids which are, of course, applied at much lower rates. Even so this switch to pyrethroids accounts for only part of the decline in quantity of active ingredient used; the introduction of economic threshold levels (ETLs) as a basis for spraying has been a major contributory factor.

All pesticide application is by back pack sprayer. Because of the manual work involved and the relatively large quantities of water needed in this type of application, ultra-low-volume sprayers (the Micron Ulva 8) and the Electrodyne electrodynamic sprayer have been evaluated and have been found to give satisfactory control.

Biological control

Parasitoids. Four species of the egg parasitoid *Trichogramma,* obtained from the Biological Control Station at Antibes in France, were evaluated for their effectiveness in controlling Lepidopterous pests on cotton. Under controlled conditions in the laboratory they showed promise. One of the species, *T. evanescens* (Westw.) (Hymenoptera: Trichogrammatidae) is also native to Morocco. This parasitoid was bred in the laboratory and released in cotton fields at the rate of 100 000 parasitoids per hectare. The results were disappointing; parasitism for spiny bollworm eggs reached only 8 percent and for leafworm only 3 percent and these rates were attained only in the immediate vicinity of the release points. It was concluded that augmentative releases of *Trichogramma* were likely to be of only limited value at times when bollworm and leafworm populations were highest because at those times insecticides were still needed to control sucking pests, especially whitefly and aphid. However, there may be a role for parasitoids released early in the season.

Predators. Among the predators of cotton pests in Morocco are *Coccinella septempunctata* L. (Coleoptera: Coccinellidae) and *Chrysoperla carnea* Stephens (Neuroptera: Chrysopidae). Predators are found in cotton fields and exercise some control of pests, especially in the late season when pesticide applications have ceased.

Indirect control

Weeds. Good weed control in cotton removes not only the crop's competitors but also the alternative host plants of many pests and diseases. Weed-free cotton also sets its crop earlier and so the season is shorter with less time for the build-up of pest populations. Trifluralin and fluometuron have given good results in experiments.

Defoliation. Early harvesting is important, not only to reduce the impact of pest attack, but to avoid early autumn rainfall and the possibility of the sudden onset of low temperatures causing premature boll split and immature fibre. Defoliants have been tested experimentally as one method of finishing the crop early but at present they are not recommended for use in Morocco where the disadvantages of their use are considered to outweigh the advantages, especially the risk of stimulating premature boll split with consequent immature fibre.

THE IMPACT OF IPM

In the Tadla region, the main cotton-growing area of Morocco, the introduction of the IPM approach to cotton pest control, especially the use of ETLs for spraying decisions, has had a very positive impact on cotton production. Improvements comprise:

- reduction in the number of spray applications from an average of 12 before 1970 to a current average of four;
- reduction in the share of crop protection as a percentage of total production costs, from 23 to 7 percent;
- reduction in the rate of application of pesticide active ingredients;
- increase in yields of seed cotton, which have doubled in recent years.

KEY PESTS

The main pests of cotton in Morocco are: spiny bollworm *(Earias insulana)*, pink bollworm *(Pectinophora gossypiella)*, Egyptian cotton leafworm *(Spodoptera littoralis)*, aphid *(Aphis gossypii)* and whitefly *(Bemisia tabaci)*.

References

El Jadd, L., Guirrou, Z., Hamdi, M. & Fahmy A. 1990. Dynamique des populations des lépidoptères ennemis du cotonnier dans les conditions du Tadla. *Rapport Convention INRA/COMAPRA,* Doc. 5, p. 13-16.

El Jadd, L., Guirrou, Z. & Semlali, M. 1990. Essais de produits chimiques en parcelles intermédiaires contre les principaux lépidoptères du cotonnier. *Rapport Convention INRA/COMAPRA,* Doc. 5, p. 17-23.

El Jadd, L., Guirrou, Z. & Ait El Alia, M. 1990a. Etude bioécologique de la mouche blanche. *Rapport Convention INRA/COMAPRA,* Doc. 5, p. 24 -35

El Jadd, L., Guirrou, Z. & Ait El Alia, M. 1990b. Essai de produits insecticides en parcelles intermédiaires contre la mouche blanche du cotonnier *B. tabaci. Rapport Convention INRA/COMAPRA,* Doc. 5, p. 36-41.

El Jadd, L., Guirrou, Z., Sekkat, A. & El Habi, M. 1990. Etude de la dynamique des populations des pucerons ravageurs du cotonnier. *Rapport Convention INRA/COMAPRA,* Doc. 5, p. 42-70.

El Jadd, L., Guirrou, Z. & Hamdi, M. 1990. Essai sur parcelles intermédiaires de produits chimiques contre les Acariens du cotonnier. *Rapport Convention INRA/COMAPRA,* Doc. 5, p. 71-73.

El Jadd, L., *et al.* 1991. Dynamique des populations des lépidoptères du cotonnier dans les conditions du Tadla durant la campagne 1989-1990.*Rapport Convention INRA/COMAPRA,* Doc. 6, p. 32-34.

El Jadd, L. & Guirrou, Z. 1991. Données bioécologiques de la mouche blanche *B. tabaci* sur cotonnier et sur plantes adventices. *Rapport Convention INRA/ COMAPRA,* Doc. 6, p. 35-44.

El Jadd, L., Guirrou, Z., Sekkat, A. & El Habi, M. 1991. Etude de la dynamique des populations d'*A.gossypii. Rapport Convention INRA/COMAPRA,* Doc. 6, p. 45-61.

ICAC. 1993. *Survey of cotton production practices 1993.* International Cotton Advisory Committee.

Pakistan

Zahoor Ahmad

INTRODUCTION

Cotton in the Indus Valley has a very long history. The oldest cotton yarn and fabric known in the world were recovered from archaeological excavations at the ancient city site of Mohenjodaro, some 320 km north of Karachi. These articles have been estimated to be about 5 000 years old. Very recently cottonseed from the fifth millennium BC has been discovered at Mehr Garh in Baluchistan.

Gossypium arboreum, an Asiatic cotton indigenous to Pakistan, was the main species grown before the introduction of the American upland cottons, *G. hirsutum,* in 1853. Originally the two species were grown as mixtures because upland cotton alone was not economic. In 1902 the Punjab Agricultural Department was established and work on the introduction of upland cottons was strengthened. A selection programme began in 1908 at the Agricultural College, Lyallpur (now Faisalabad). The first variety to come from the programme, 3F, was found to be highly susceptible to jassid attack. A second variety, 4F, which was introduced in 1914, had a hairy leaf character that conferred resistance to jassid. This variety proved commercially successful. Since 1914, varieties of American upland origin have gradually replaced the indigenous cottons and now account for 97 percent of the total area under cotton in Pakistan.

The balance between the export of raw cotton and local consumption has changed over time, as the local textile industry has developed and the international market diversified. Local mills used 8 159 000 bales in 1993, compared with 1 153 000 bales in 1961, and exports have come to be dominated by yarn and piece goods (by value yarn accounts for 23 percent of cotton exports, fabrics 19 percent, garments 14 percent, bedding 10 percent and knitwear and hosiery 12 percent).

About 400 000 tonnes of cottonseed oil is produced annually, accounting
for 85 percent of the country's edible oil production.

Pakistan has become one of the world's largest cotton producers, now
ranking fourth in both area and production. Pakistan exports of yarn account
for 29 percent of the total world export market. Revenue from cotton, in the
form of export duties and other taxes, is a major source of income for the
government and cotton is also the main earner of foreign currency.

AREA, PRODUCTION AND YIELD

Table 10 shows the area under cotton, total production and yield of lint for
selected years over the period 1947/48 to 1993/94. Production peaked at
12.821 million bales in the 1991/92 season, when the area also reached

TABLE 10
Cotton area, production and yield in Pakistan

Season	Area ('000 ha)	Production ('000 bales)[1]	Yield (kg lint per ha)
1947/48	1 237	1 156	159
1950/51	1 221	1 406	295
1960/61	1 293	1 692	233
1970/71	1 733	3 050	313
1980/81	2 108	4 201	339
1985/86	2 364	7 154	515
1986/87	2 505	7 760	527
1987/88	2 568	8 633	572
1988/89	2 620	8 385	544
1989/90	2 599	8 551	559
1990/91	2 700	9 600	602
1991/92	2 835	12 821	768
1992/93	2 835	9 053	543
1993/94	2 803	7 935	481

Source: Pakistan Central Cotton Committee, Ministry of Agriculture, Government of Pakistan.
[1] 1 bale = 170 kg.

a peak at 2 835 000 ha. Since the 1991/92 season production has declined and Pakistan has been forced to import raw cotton to sustain the local textile industry. It has lost its long-held second place as an exporter of raw cotton.

The decline in production was mainly caused by a decline in yields which dropped from 768 kg of lint per hectare in 1991/92 to 481 kg of lint per hectare in 1993/94. The area under cotton did not fall much in this period. Although the lint yields have increased at least fourfold since 1947/48, they are still low compared to those of irrigated cotton in countries such as Australia and the United States.

Because of the importance of cotton to the national economy, the Government of Pakistan has made considerable efforts to increase output, especially over the past five years. Measures taken have included the provision of price incentives and credit facilities on liberal terms, the subsidizing of inputs including agricultural machinery, the introduction of pest monitoring schemes and mass education and training schemes for farmers. These, and other changes, have dramatically altered the socio-economic context of cotton farming. Some of the more important features of the cotton production system may be summarized as follows:

- The summer weather in the cotton-growing areas is very hot and dry and there is insufficient water to grow the other main summer cash crop, sugar cane. Farmers therefore have little choice but to grow cotton and, being for the most part resource-poor, need to maximize their returns from the crop.

- The government determines the minimum price farmers will receive for their seed cotton. The Cotton Export Corporation (CEC), a public-sector organization, ensures that cash is available to pay farmers on time for their cotton by undertaking to purchase lint from the ginneries. This price support system has encouraged farmers to grow cotton and to invest in modern production methods to maximize yields. Since the 1988/89 season the private sector has been allowed to enter the cotton market and this has increased competition between buyers.

- Pesticides are supplied to farmers by the private sector, usually on the basis of six months interest-free credit. Pesticide use increased from 905 tonnes active ingredient (ai) in 1981 to 4 919 tonnes ai in 1993.

Pesticides are used in integrated pest management (IPM) programmes, which ensures that they are used rationally and profitably.

- Farmers can obtain credit from government sources through cooperative societies, commercial banks and the Agricultural Development Bank, for the purchase of inputs such as seed, fertilizer and pesticides. Other sources of credit include landlords, village shopkeepers, cotton buyers and agricultural supply dealers. Ginneries may offer loans in kind.

- Nearly all cotton farmers are now mechanized, with tractor power having almost completely replaced draft animals for cultivation over the last five years. Those farmers who do not own their own tractors hire them. Many farmers have installed their own tubewells to supplement canal irrigation and the use of fertilizers on cotton is now common practice.

- The increased profitability of cotton has enabled farmers to invest in other crops in the rotation, thereby increasing the intensity of land use, which is now up to 157 percent. Maize, wheat and sunflower are increasingly used as break crops between cotton crops, replacing the former fallow period.

- Many large and progressive farmers have invested in the infrastructure supporting cotton production, including ginneries and seed companies, thereby increasing their profits and providing rural employment. There are now 50 private seed companies selling certified seed in competition with public-sector companies; ten years ago there were none.

- There is now an element of competition between cotton farmers, which has the beneficial effect of ensuring that production costs are minimized through the efficient use of inputs while profitability and yields are increased. Many progressive farmers have established direct links with research institutes to ensure that they are kept informed of developments in production technology.

- The increase in cotton production in Pakistan has encouraged investment in the textile industry. Since 1981 some 200 new mills have been opened, at a rate of about 12 new mills each year. Local consumption of raw cotton has increased from 2.7 million bales in 1982 to 8.2 million bales in 1994. This expansion of the textile industry has created thousands of new jobs, particularly in the urban areas.

CULTURAL PRACTICES
Land preparation
About 70 percent of the cotton crop follows wheat, the rest follows sunflower, maize, soybean, pulses or fallow. Land preparation depends on the previous crop; four to five ploughings and one to two plankings are usually sufficient following wheat harvest. Land levelling is of great importance to ensure even water distribution under the predominant system of basin irrigation. If wheat has been harvested by machine the straw is disposed of, either by deep burial with a mould-board plough, after chopping with a disc-harrow or rotavator, or by burning. Other crop residues are similarly chopped and incorporated or else removed from the field.

Where cotton follows wheat or fallow, two heavy presowing irrigations are applied and the land is ploughed between the two irrigations to ensure the deep penetration of water and thus reduce the number of postsowing irrigations needed. After the irrigations, when the land becomes workable, it is ploughed to a depth of 15 to 20 cm. On average three to four cross-cultivations, each followed by planking to break down clods, are sufficient to prepare a seed-bed for cotton. On large, mechanized farms, rotavators, disc-harrows or spring-tined harrows with crumblers are also used for these presowing cultivations.

Crop management
Cotton is sown between 15 April and 15 May in Sind and between 15 May and 15 June in Punjab. Most cotton is sown by machine, usually four rows at a time at seed rates of 8 to 10 kg acid delinted seed per hectare. Inter-row spacing is 75 cm and intrarow spacing varies from 20 to 30 cm. This spacing is reduced to 15 to 20 cm for late-sown cotton. The recommended plant population is 45 000 to 50 000 plants per hectare. Cotton is normally sown on the flat and may be ridged up before the canopy closes over. On clay and saline soils cotton is sown on ridges, dibbled into the side of the ridge on saline soils.

Germination is complete within seven days, and gap filling is carried out over the following five days to ensure germination on residual moisture. The first thinning is completed within three to four weeks of germination and thinning is finally completed by the time of the first postsowing irrigation. In areas where the black-headed cricket causes serious losses of seedlings thinning is delayed.

Weeds are controlled by one or two dry hoeings with two or three more following irrigation.

Fertilizer

Cotton is somewhat less demanding in its nutrient requirements than maize, wheat or sugar cane and considerably less so than vegetables. In Pakistan, excessive nitrogen can lead to problems with rank growth. The response of cotton to phosphorus (P) in Pakistan is not consistent, although requirements would seem to be very low. It is recommended that cotton following fertilized wheat does not receive any additional P, but following sunflower or maize phosphatic fertilizers should be applied at the rate of 45 to 65 kg per hectare. Pakistan soils are generally rich in potash (K) and the application of K does not increase yields. There has been some response to boron and sulphur and research is currently being undertaken on these minor elements.

Irrigation

The timing of the first postsowing irrigation depends on the cotton variety being grown. For the varieties NIAB-78, S-12, CIM-109 and CIM-240 the first irrigation is applied 30 to 35 days after sowing and subsequent irrigations are at 12- to 15-day intervals. For the varieties MNH-93, BH-36 and MNH-147 the first irrigation is 45 to 50 days after sowing.

Plant growth accelerates once flowering begins and water requirements reach a maximum. It is essential that the crop should not suffer from any water stress during the period of boll formation and maturation if fruiting point shedding and restricted boll development are to be avoided. Only in areas where the water-table is high can the interval between irrigations during the fruiting phase be extended to 20 days or longer.

The timing of the final irrigation is important for both economy of water use and final yield. There is less need for water during the period of boll opening and only sufficient to allow late formed bolls to mature properly is needed. In Punjab the last irrigation is applied at the end of September or in early October. Late irrigations delay maturation of the crop and encourage boll rots and the persistence of pink bollworm infestations. There is seldom any benefit from late irrigations.

COTTON VARIETIES

A large number of *Gossypium hirsutum* varieties have been bred in Pakistan to suit local conditions and to meet market demands for particular fibre characteristics. There are now short- and long-season, early- and late-maturing, heat-tolerant and high-yielding varieties. In recent years there has been considerable improvement in ginning out-turn and staple length and Pakistan cottons are noted for their high maturity and fibre strength. Table 11 lists the main commercial varieties, together with their year of release and main fibre characteristics. The agronomic features of the more important commercial varieties currently under cultivation are discussed in the following.

MNH-93. This variety was developed at the Cotton Research Station, Multan, through hybridization. It is of medium height and large bolled. It has a tendency to lodge when bearing a full load of fruit. Its seed is fuzzy white. It is profusely hairy and fairly tolerant to the attack of jassid. It has a yield potential of 3 500 kg of seed cotton per hectare. It was approved for general cultivation in 1980. It is grown on about 15 percent of the total cotton area in Punjab.

NIAB-78. This variety was developed at the Nuclear Institute for Agriculture and Biology, Faisalabad, by hybridization followed by irradiation. It is early-maturing and high-yielding (up to 4 500 kg seed cotton per hectare). It is of medium height, erect, with a sympodial branching habit. It has a medium-sized boll with fuzzy white seed. It is less tolerant to the attack of jassid than MNH-93. It is grown on about 35 percent of the total cotton area in Punjab.

MS-84. This variety was developed at the Cotton Research Station, Multan, through hybridization. It is a long staple with a staple length of 1 1/4 inches

TABLE 11

Fibre characteristics of commercial varieties of cotton in Punjab and Sind provinces

Variety	Year of release	Ginning out-turn (%)	Staple length (inches)	Micronaire	Fibre strength (tppsi)
Punjab upland					
(G. hirsutum)					
B-557	1975	34.5	33/32	4.5	92.5
MNH-93	1980	36.0	17/16	4.6	94.0
NIAB-78	1983	35.5	33/32	4.7	92.0
MS-84	1983	33.0	5/4	4.2	91.3
SLH-41	1985	36.0	33/32	4.4	95.8
MNH-129	1986	36.5	17/16	4.6	95.4
CIM-70	1986	31.5	9/8	4.2	92.5
S-12	1988	40.0	17/16	4.8	92.0
FH-87	1988	34.0	33/32	4.5	92.0
CIM-109	1990	35.1	17/16	4.4	92.0
GOHAR-87	1990	34.5	17/16	4.4	92.6
RH-1	1990	31.5	9/8	3.9	103.7
NIAB-86	1990	34.5	9/8	4.2	94.0
CIM-240	1992	36.5	17/16	4.7	93.7
MNH-147	1992	41.3	9/8	4.2	95.5
FH-682	1992	37.0	9/8	4.3	95.7
BH-36	1992	38.7	35/32	4.3	100.3
NIAB-26N	1992	37.5	35/32	4.4	95.0
Desi					
(G. arboreum)					
D-9	1971	39.0	5/8	7.5	80.0
Ravi	1982	41.0	5/8	8.0	80.0
Rohi	1986	38.8	5/8	8.0	80.0
Sind upland					
(G. hirsutum)					
Qallandri	1974	34.0	1 1/8	3.5	90.0
S-59-1	1975	34.0	1 1/8	3.5	92.5
K-68-9	1977	33.0	1 3/16	4.3	96.0
Rehmani	1985	35.0	1 1/16	4.4	90.0
Shaheen	1988	34.0	1 1/16	4.2	95.0
Reshmi	1991	36.0	1 1/4	4.1	97.0
CRIS-9	1992				
Desi					
(G. arboreum)					
SKD-10/19	1976	40.0	5/8	9.5	-

(32 mm). The plants are of medium height with fairly large bolls. The yield potential is about 2 500 kg of seed cotton per hectare. Its seed is bold fuzzy white. Being hairy it is fairly tolerant to jassid. It was approved for general cultivation in 1983.

S-12. This variety was developed at the Cotton Research Station, Multan, through hybridization. It is a short stature, early-maturing and high-yielding variety (4 500 kg seed cotton per hectare). It possesses a high ginning out-turn (40 percent). It is a monopodial plant and bears six to eight monopodia. It has a low tolerance to jassid. It was released for cultivation in 1988. It formerly occupied 45 percent of the cotton area in Punjab but because of susceptibility to leaf curl virus the area had declined to about 16 percent by the 1993/94 season.

CIM-109. This variety was developed at the Cotton Research Institute, Multan, through hybridization. It is a high-yielding variety with a potential of 4 500 kg of seed cotton per hectare. It is early-maturing and so is suitable for a cotton-wheat rotation. It was approved for general cultivation in 1990 in Multan, Bahawalpur, DG Khan and Faisalabad Divisions. It is of medium height, erect, with a sympodial branching habit. It is fairly tolerant to jassid and to stunting. It is also tolerant to leaf curl virus. It is suitable for early sowing in May.

CIM-29. This variety was developed through the hybridization of a local commercial variety, CIM-70, and an exotic Australian variety, W-1106. It was approved for general cultivation in 1992. Its plant is of medium height with a sympodial branching habit. It has large bolls with fluffy openings and is very easy to pick. It is fairly tolerant to leaf curl virus in comparison with other commercial varieties. It is also early-maturing and suitable for a cotton-wheat rotation. It has a yield potential of 5 000 kg of seed cotton per hectare. It was grown on 17 percent of the cotton area in the 1993/94 season in Punjab.

BH-36. This variety was developed at the Cotton Research Station, Bahawalpur, through the hybridization of BS-1 with Tx Bonham-76C. It is a tall variety with medium boll size. It has a good yield potential and is tolerant of leaf curl virus and jassid.

MNH-147. This variety was developed at the Cotton Research Station, Multan, through the hybridization of 431/79 and 283/80. It was approved for general cultivation in 1992. It has the yield potential of 3 500 kg per hectare. Its ginning out-turn is 41.4 percent and staple length is 1 1/8 inches (28.8 mm). It is a tall and late-maturing variety, fairly tolerant to leaf curl virus.

K-68-9. This variety was developed at the Cotton Research Station, Ghotki, through the hybridization of 199F and Wilds in 1977. It is of medium plant height with large bolls and is late-maturing. Its seed is fuzzy white. It is fairly tolerant to jassid and has a yield potential of 3 000 kg of seed cotton per hectare.

Rehmani. This variety was developed at the Cotton Research Institute, Tandojam, through the hybridization of *G. hirsutum*-21 and McNaire-TH-14920. It was released for general cultivation in 1985. Its yield potential is about 3 500 kg of seed cotton per hectare and it is hairy and tolerant to the attack of jassid.

Shaheen. This variety was developed at the Cotton Research Institute, Ghotki, through hybridization of GH-7/72 and H-1. It is a high-yielding variety with a potential of about 3 500 kg per hectare. It is of medium height and fairly tolerant to jassid.

CRIS-9. This variety was developed at the Cotton Research Institute, Sakrand, by hybridization of RA47-33 and Rajhans and was approved for cultivation in Sind in 1992. It is an early-maturing, sympodial plant, with a good yield potential of 4 500 kg seed cotton per hectare.

PESTS
Although cotton is subject to attack by a number of pests and diseases and competition from many weed species, only a few are of economic importance.

Losses to pests vary considerably, but are usually within the range of 20 to 40 percent; in 1993 pest attack was particularly severe and losses were estimated at 40 percent.

Insects and mites

Scirtothrips dorsalis *Hood and* **Thrips tabaci** *(Lindeman) (Thysanoptera: Thripidae).* Thrips are distributed throughout Pakistan. Early-sown cotton is attacked at the seedling stage; later attacks give cotton plants a characteristic blackish colour, although sooty mould is not present.

Aphis gossypii *(Glov.) (Hemiptera: Aphididae).* Aphid is now one of the key pests of cotton, having increased in importance over the past few years. Aphid feeding reduces plant vigour, resulting in lower yields, while honeydew excretion reduces lint quality through stickiness and the buildup of sooty mould. Aphid has many predators, including coccinellid beetles, and parasitoids in Pakistan and chemical control should be undertaken only after careful monitoring in order to preserve, if possible, the beneficial species (Ahmad, Attique and Rashid, 1985).

Amrasca devastans *(Distant) (Hemiptera: Cicadellidae).* Jassid is the most important of the sucking pests and is considered a key pest of cotton throughout the cotton areas. Feeding by nymphs and adults, mainly on the lower surface of the leaf, causes characteristic "hopper burn" and affected leaves may be shed prematurely. The most serious infestations occur between July and September and are favoured by high humidity; breeding is inhibited by dry conditions. Varieties with leaf hair show resistance to jassid, which is also controlled chemically with granular formulations of insecticide.

Bemisia tabaci *(Gen.) (Hemiptera: Aleyrodidae).* Whitefly has recently become a key pest in Pakistan, mainly because of its importance as a vector of leaf curl virus, which was responsible for the decrease in production in the 1992/93 and 1993/94 seasons, but also because of the direct damage its feeding does as well as the loss of lint quality caused by honeydew stickiness and the buildup of sooty mould. Small numbers of whitefly may be present on cotton throughout the season, but continuous dry conditions favour the pest and populations may increase rapidly to damaging levels. The eradication of weeds in cotton fields discourages whitefly. The pest is resistant to various types of insecticides, so chemical control is becoming difficult to achieve.

Polyphagotarsonemus latus *(Banks) (Acari: Tarsonemidae).* Yellow tea mite is a polyphagous pest that can become prevalent under conditions of high humidity. Adults and immature stages feed on the lower surface of the leaf, which initially appears smooth and shiny and later becomes dark green with the leaf edges curled down. Leaves then tear, giving a ragged appearance. The loss of photosynthetic tissue retards plant development and fruiting points may be shed.

Tetranychus *spp. (Acari: Tetranychidae).* Spider mites are minor pests of cotton in Pakistan, usually appearing late in the season. Heavy infestations may cause leaf shedding and induce premature boll opening. Careful selection of insecticides can help to ensure that spider mite problems do not develop.

Pectinophora gossypiella *(Saund.) (Lepidoptera: Gelechiidae).* Pink bollworm is the most serious bollworm pest throughout the cotton areas, but because of the longer and later growing season in the Punjab it is more important there than in Sind where the earlier harvest stops the buildup of populations within the cotton crop.

The presence of pink bollworm larvae in fruiting points is not easily detected even when bolls are cut open. The presence of larvae within flowers causes the characteristic rosette effect, as silk threads spun by the larvae prevent the flower from opening properly. Damage later in the season as populations build up is more serious than are early-season attacks. In Punjab most damage is caused in September and October. When bolls are attacked, yield is lost and fibre quality is reduced, as is the value of the seed for oil production.

The very high temperatures often experienced in Pakistan in summer can result in sterility in female pink bollworm moths and it is not until temperatures drop below 35°C that populations start to increase, the increase coinciding with the main period of boll formation.

Carryover from one season to the next is as diapause larvae in left-over bolls on cotton stalks stored to provide fuel. This is the weak link in the annual cycle of pink bollworm populations. Destruction of these bolls effectively reduces the carryover population; it can be achieved by allowing stock to graze cotton fields after harvest and before stalks are cut down. A new component of the pink bollworm control strategy is the introduction of early-maturing varieties, including CIM-70, NIAB-78, CIM-109 and CIM-240, to Punjab, allowing the harvest there to be completed by the end of September, as in Sind. Other control methods being evaluated include the use of pheromones and the breeding of nectariless cottons.

Earias insulana *(Boisd.) and* **E. vittella** *(F.) (Lepidoptera: Noctuidae).* Spotted and spiny bollworms are the other main bollworm species. *E. vitella* predominates, but both species are found throughout the cotton areas and infest other malvaceous species as well as cotton. Okra, *Hibiscus esculentus* L., in particular provides an alternate host in the non-cotton season and infestations on cotton have been found to be heaviest in fields next to okra crops. Farmers are advised to finish okra cultivation by the end of April or at the latest mid-May and to plough in crop residues to destroy the bollworms. This practice can be very effective in reducing spotted bollworm attacks.

Early in the cotton season, before fruiting bodies are available, spotted bollworm acts as a stem borer, entering the terminal bud of a vegetative shoot and boring downwards from the growing point, or it may enter directly into an internode. Stem boring does not reduce yield but may delay maturation of the crop. Later the attack is switched to buds, flowers and immature bolls.

Helicoverpa armigera *(Hb.) (Lepidoptera: Noctuidae).* American bollworm is generally considered a minor pest of cotton, maize and other crops in

Pakistan, but over the years the infestation of cotton has shown an increase, possibly because of indiscriminate use of broad-spectrum insecticides.

Spodoptera litura *(F.) (Lepidoptera: Noctuidae).* Leafworm is usually a minor pest of cotton in Pakistan but occasionally an outbreak occurs which can cause complete defoliation.

Microtermes *spp. (Isoptera: Termitidae).* Termites are of little economic importance on cotton. The application of fresh farmyard manure encourages attacks, which usually occur early in the season. Stems are damaged or severed below ground, causing the plant to die.

Gryllus bimaculatus *(De Geer) (Orthoptera: Gryllidae).* Crickets, both adults and nymphs, attack cottonseed following sowing and prevent germination. Young seedlings may also be attacked. Crickets, of which there are several species with *G. bimaculatus* being the most common, are restricted to clay soils.

Diseases

Cotton leaf curl virus. Cotton leaf curl virus disease was considered of only minor significance until 1988 when some 60 ha of cotton were affected in Punjab. Since then the area affected has increased each year and by 1992 covered 0.5 million ha, resulting in an estimated yield loss of 750 000 bales. By 1993 the disease had spread to all the cotton-growing areas of Punjab, although the intensity of attack varied, and it was noted in both Sind and the Dera Ismail Khan area of the Northwest Frontier Province. Yield losses were put at 1.9 million bales (Table 12).

Cotton leaf curl virus is now the most important cotton disease in Pakistan. The cause is thought to be a whitefly-transmitted Gemini virus, although more research is needed to confirm this. Symptoms of the disease include either upward or downward curling of the leaves associated with thickening of the veins, especially the small veins. In extreme cases the formation of enations occurs on the underside of the leaf. Yield is considerably reduced.

The long-term control strategy is to breed resistant varieties, meanwhile it is recommended that high standards of crop husbandry should be maintained

TABLE 12

The incidence of leaf curl virus disease 1988/89 to 1993/94

Year	Affected area ('000 ha)			Loss of production ('000 bales)
	Partial	Complete	Total	
1988/89	-	0.06	0.06	0.3
1989/90	-	0.2	0.2	1.0
1990/91	-	0.8	0.8	4.0
1991/92	11.3	2.8	14.1	20.0
1992/93	364.0	121.0	485.0	750.0
1993/94	607.0	282.0	889.0	1 880.0

and only varieties showing tolerance to the disease should be grown. Reduction of whitefly populations through management or eradication of alternate hosts should be practised.

Bacterial blight. *Xanthomonas campestris* pv *malvacearum* (E.F. Smith) Dye is a disease of considerable complexity and causes heavy loss of yield in cotton in Pakistan.

Seed and seedling diseases. Both soil-borne and seed-borne pathogens cause losses of stand and plant vigour in Pakistan.

Root rots. No certain causal organism of root rots has been identified, although the pathogens most frequently associated with these rots include *Rhizoctonia solani* Kuhn, *Macrophomina phaseolina* (Tassi) Goid. (syn. *R. bataticola*) and *Fusarium* sp. Root rot problems usually appear in patches in a field and remain confined to these patches year after year.

Stunt disease. The cause of stunt, which may affect the plant at any stage, is not known, but it could be a physiological disorder or be of viral origin. The disease was particularly common during the early introductions of varieties of American origin.

Boll rots. Boll rots can cause heavy loss of yield. There are two types of causal organisms, those capable of invading the boll tissue on their own – primary boll rots – and those which invade through wounds caused by insects, other diseases or physical factors – secondary boll rots. Examples of the first group of pathogens include *Botryodiplodia theobromae* Pat., *(syn. Diplodia gossypina), Colletotrichum capsici* (Syd.) Butler & Bisby, *Myrothecium roridum* Tode ex Fr. and *Xanthomonas campestris* pv *malvacearum* (E.F. Smith) Dye. The second group includes *Aspergillus* spp., *Alternaria* spp., *Rhizopus* spp. and many other weak pathogens.

Weeds

Trianthema portulacastrum *L. (Aizoaceae).* A serious perennial weed favoured by soils rich in organic matter, especially farmyard manure, this plant dies down in winter and sprouts again in spring.

Cyperus rotundus *L. (Cyperaceae).* This weed is tolerant of a wide range of soil types, but waterlogged depressions in fields are particularly prone to infestation. Line sowing of cotton appears to favour the weed and it is very difficult to control by hoeing and weeding.

Cynodon dactylon *(L.) Pers. (Graminae).* A perennial grass weed propagated by the spread of underground stems, eventually it forms a dense mat which covers the soil completely. Mechanical control tends to increase the spread of the weed. Its nutritive value is high, making it a good source of fodder.

Sorghum halepense *(L.) Pers. (Graminae).* Like the previous species this perennial grass weed also spreads by underground stems (rhizomes), although the seeds are another important means of reproduction. It grows on almost all types of soil and plants may reach a height of 75 to 150 cm.

Convolvulus arvensis *L. (Convolvulaceae).* A perennial climber, this weed spreads by root runs and seed. Each portion of the root is capable of producing a new plant, so mechanical control is difficult.

Tribulus terrestris *L. (Zygophyllaceae).* The seeds of this fast-growing hardy weed are protected by a hard covering which keeps them viable for a long time. The shoots creep along the ground.

Amaranthus *spp. (Amaranthaceae).* An annual weed, it produces a tender plant of up to 30 cm in height and is propagated by seed.

Euphorbia prostrata *Ait. (Euphorbiaceae).* This is an annual weed with a characteristic milky sap.

CONTROL METHODS
Chemical control
Pesticide use in Pakistan is dominated by insecticides applied to cotton, which account for about 75 percent of the pesticide market. The organophosphates and synthetic pyrethroids are the most important groups in terms of both value and quantity; use of carbamates and organochlorines is declining. Table 13 shows the growth in private-sector pesticide imports since 1981, which exceeded 500 percent by 1993. The use of herbicides and fungicides is low in comparison with that of insecticides; most herbicides are used in cotton, wheat and potatoes, while vegetables are the main market for fungicides which are not generally used in cotton. A complete list of all the insecticides and insecticide combinations registered for use on cotton, together with rates of application against the main target pests, is given in Table 14.

The Government of Pakistan withdrew from pesticide importation and sale in 1980 and handed responsibility for these matters over to the private sector. The federal government remains responsible for pesticide registration, regulation and the issue of import licences. Farmers are normally responsible for applying pesticides, but the Federal Plant Protection Department maintains a fleet of aircraft to combat pest outbreaks, usually of locusts, although aerial spraying is occasionally used on sugar cane and rice, when farmers must pay for the insecticides used (operational costs being paid for by provincial governments). Otherwise government involvement in pesticide application is now negligible.

TABLE 13

**Quantity of pesticides imported by the private sector
in Pakistan**

Year	Quantity *(tonnes active ingredient)*	Growth per year *(%)*	Growth since 1981 *(%)*
1981/82	905	-	-
1982/83	1 345	49	49
1983/84	1 757	31	94
1984/85	2 585	47	185
1985/86	3 489	35	285
1986/87	4 111	18	354
1987/88	4 429	8	389
1988/89	4 065	8	349
1989/90	4 706	16	420
1990/91	5 730	22	533
1991/92	5 920	3	554
1992/93	5 619	-5	520
1993/94	4 911	-13	443

The multinational agrochemical companies supplement the activities of government extension services by running training programmes for pesticide dealers and farmers and holding group meetings of between 50 and 100 farmers. These activities promote the products of the agrochemical companies and teach modern cotton production methods to the farmers. Some village-level pesticide dealers provide services for cotton farmers, hiring out sprayers and labour for spraying or providing a contract spraying service.

Integrated Pest Management (IPM) is the basis for the use of insecticides on cotton. So far use of insecticides has had most impact in Punjab where there has been provincial government support for IPM. Figure 1 shows the changes in levels of infestation by key pests over the years 1983 to 1993. Infestations, as represented by so-called "hot spots", have generally declined over the period as insecticide use has spread to increasing numbers of farmers.

TABLE 14

Insecticides and insecticide mixtures registered for use on cotton, 1991

Common name	Formula-tion[1]	Type	Mode of action	Mammalian toxicity	Target pests	Applica-tion per acre[2]
Acephate	75 sp	Organophosphorus	Systemic, stomach	Low	Bollworms	700 g
Aldicarb	15 gr	Carbamate	Systemic, contact	Very high	Sucking pests,	3.5 kg
Alpha-cypermethrin	5 ec	Pyrethroid	Contact, stomach	Medium	Bollworms	200 ml
Amitraz	20 ec	Amidine	Contact, fumigant	Medium	Sucking pests, bollworms and mites	1 000
Azinphos	20 ec	Organophosphorus	Contact, stomach	High	Pink bollworm	1 750 ml
Beta-cyfluthrin	2.5 ec	Pyrethroid	Contact, stomach	High	Bollworms	200 ml
Bifenthrin	10 ec	Pyrethroid	Contact, stomach	High	Sucking pests, bollworms	250 ml
Binapacryl	40 ec	Dinitrophenyl butenoate	Contact	Medium	Mites	1 000 ml
Bromophos	25 ec	Organophosphorus	Contact, stomach	Low	Broad spectrum	500 ml
Bromophos ethyl	80 ec	Organophosphorus	Contact, stomach	High	Broad spectrum	500 ml
Bromopropylate	50 ec	Benzilate	Contact	Very low	Mites	500 ml
Carbaryl	85 wp	Carbamate	Contact	Medium	Insect complex	1.4 kg
Carbosulfan	20 ec	Carbamate	Systemic, contact, stomach	Medium	Sucking pests	500 ml
Chlorfluazuron	5 ec	Benzoylurea	csi[3]	Low	Bollworms	400 ml
Chlorobenzilate	50 ec	Benzilate	Contact	Low	Mites	500 ml
Chlorpyrifos	40 ec	Organophosphorus	Contact, stomach	Medium	Sucking pests, bollworms	1 000 ml
Chlorthiophos	50 ec	Organophosporus	Contact	High	Broad spectrum	1 000 ml
Cycloprothrin	2 ec	Pyrethroid	Contact	Medium	Bollworms, jassid	400 ml
Cyfluthrin	5 ec	Pyrethroid	Contact, stomach	Medium	Bollworms	250 ml
Cyhexatin	50 wp	Organotin	Contact	Low	Mites	740 g
Cypermethrin	10 ec	Pyrethroid	Contact, stomach	Medium	Bollworms	250 ml
Deltamethrin	2.5 ec	Pyrethroid	Contact, stomach	Medium	Bollworms	250 ml
Demeton-s-methyl	25 ec	Organophosphorus	Contact, systemic	High	Sucking pests	750 ml
Dichlorvos	100 ec	Organophosphorus	Contact, stomach	High	Broad spectrum	-
Dicofol	18.5 ec	Organochlorine	Contact	Low	Mites	2 000 ml
Dicrotophos	87 sl	Organophosphorus	Systemic	High	Jassid, thrips, sucking pests	-
Dieldrin	20 ec	Organochlorine	Contact, stomach	High	Cutworms, termites, crickets, grasshoppers	2 500 ml
Dimethoate	40 ec	Organophosphorus	Contact, systemic	Medium	Sucking pests	500 ml
Disulfoton	10 gr	Organophosphorus	Systemic	Very high	Sucking pests, mites	5 kg
Endosulphan	35 ec	Organochlorine	Contact, stomach	Medium	Bollworms	1 250 ml
Esfenvalerate	0.5 ul	Pyrethroid	Contact, stomach	Medium	Bollworms, jassid	2 500 ml
Ethion	46.5 ec	Organo-phosphorus	Contact, stomach	Medium	Sucking pests	1 000 ml
Fenitrothion	50 ec	Organo-phosphorus	Contact	Medium	Bollworms and sucking pests	1 000 ml
Fenpropathrin	10 ec	Pyrethroid	Contact, stomach	Medium	Insect complex	750 ml
Fenvalerate	20 ec	Pyrethroid	Contact, stomach	Medium	Bollworms	250 ml
Flucythrinate	10 ec	Pyrethroid	Contact	High	Bollworms	25 ml
Flufenoxuron	10 sl	Benzoylurea	csi[3]	Low	Bollworms	400 ml
Formothion	25 ec	Organophosphorus	Contact, stomach	Medium	Sucking pests	1 000 ml
Gamma-HCH	10 dp	Organochlorine	Contact	Medium	Crickets	2.5 kg
Heptachlor	32 ec	Organochlorine	Contact, stomach fumigant	Medium	Termites, crickets, grasshoppers	2 250 ml
Isoxathion	50 ec	Organophosphorus	Contact,	High	Sucking pests	1 000 ml
Lambda-cyhalothrin	2.5 ec	Pyrethroid	Contact	High	Bollworms	325 ml
Malathion	57 ec	Organophosphorus	Contact, stomach	Low	Sucking pests, armyworm	750 ml
Methamidophos	600 sl	Organophosphorus	Contact, stomach	High	Sucking pests	600 ml
Methidathion	40 ec	Organophosphorus	Contact, stomach	High	Bollworms, sucking pests	1 000 ml

(cont.)

TABLE 14 (continued)

Common name	Formula-tion[1]	Type	Mode of action	Mammalian toxicity	Target pests	Applica-tion per acre[2]
Methomyl	29 sl	Carbamoy-loxime	Contact, stomach	High	Bollworms	1 000 ml
Monocrotophos	40 sl	Organophosphorus	Contact, systemic	High	Insect complex	1 000 ml
Parathion-methyl	50 ec	Organophosphorus	Contact, systemic	Very high	Sucking pests, bollworms	750 ml
Phorate	10 gr	Organophosphorus	Systemic, contact	Very high	Sucking pests	5 kg
Phosalone	35 ec	Organophosphorus	Contact	Medium	Sucking pests, bollworms	750 ml
Phosphamidon	100 ec	Organophosphorus	Systemic, stomach	Very high	Sucking pests	250 ml
Pirimiphos-ethyl	30 ec	Organophosphorus	Contact, stomach	High	Insects in or on the soil	215 ml
Pirimiphos-methyl	50 ec	Organophosphorus	Contact	Low	Sucking pests	500 ml
Profenofos	500 ec	Organophosphorus	Stomach, contact	Medium	Insect complex	800 ml
Propargite	57 ec	Sulphite acaricide	Contact, stomach	High	Mites	750 ml
Pyraclofos	50 ec	Organophosphorus	Contact, stomach	Medium	Sucking pests, bollworms	800 ml
Quinalphos	25 ec	Organophosphorus	Contact, stomach	Medium	Broad spectrum	1 000 ml
Tau-fluvalinate	20 ec	Pyrethroid	Contact, stomach	Medium	Broad spectrum	175 ml
Thiodicarb	80 df	Carbamoyloxime	Contact, stomach	Medium	Insect complex	480 ml
Thiometon	25 ec	Organophosphorus	Systemic, stomach, contact	High	Thrips only	600 ml
Tralomethrin	3.6 ec	Pyrethroid	Systemic, stomach, contact	Low	Bollworms	250 ml
Triazophos	40 ec	Organophosphorus	Contact, stomach	High	Insect complex	1 000 ml
Cyfluthrin + methamidophos	525 ec	Pyrethroid + organophosphorus	Stomach	High	Insect complex	400 ml
Cypermethrin + chlorpyrifos	505 ec	Pyrethroid + organophosphorus	Contact, stomach	Medium	Insect complex	500 ml
Deltamethrin + dimethoate	312 ec	Pyrethroid + organophosphorus	Contact, systemic	Medium	Insect complex	325 ml
Cypermethrin + profenofos	440 ec	Pyrethroid + organophosphorus	Contact, stomach	Medium	Insect complex	600 ml
Fenvalerate + dimethoate	40 ec	Pyrethroid + organophosphorus	Contact, stomach	Medium	Insect complex	350 ml
Flucythrinate + dimethoate	25 ec	Pyrethroid + organophosphorus	Contact, stomach	High	Insect complex	400 ml
Flucythrinate + mephosfolan	22.5 ec	Pyrethroid + organophosphorus	Contact, stomach	High	Insect complex	500 ml
Cypermethrin + dimethoate	25 ec	Pyrethroid + organophosphorus	Contact, stomach	Medium	Insect complex	1 000 ml
Cypermethrin + dimethoate	300 ec	Pyrethroid + organophosphorus	Contact, systemic	Medium	Insect complex	500 ml
Cypermethrin + mephosfolan	350 ec	Pyrethroid + organophosphorus	Contact, systemic	Very high	Insect complex	400 ml
Beta-cyfluthrin + methamidophos	525 sl	Pyrethroid + organophosphorus	Contact, systemic	High	Insect complex	400 ml
Cypermethrin + monocrotophos	44 ec	Pyrethroid + organophosphorus	Contact, systemic	High	Insect complex	500 ml
Cypermethrin + monocrotophos	425 ec	Pyrethroid + organophosphorus	Contact, systemic	High	Insect complex	1 000 ml
Cypermethrin + monocrotophos	425 ec	Pyrethroid + organophosphorus	Contact, systemic	High	Insect complex	600 ml
Alpha-cypermethrin + monocrotophos	42 ec	Pyrethroid + organophosphorus	Contact, systemic	High	Insect complex	500 ml
Deltamethrin + triazophos	36 ec	Pyrethroid + organophosphorus	Contact, systemic	High	Insect complex	600 ml
Cypermethrin + dimethoate	25 ec	Pyrethroid + organophosphorus	Contact, systemic	Medium	Insect complex	1 000 ml
Fenpropathrin + fenitrothion	50 ec	Pyrethroid + organophosphorus	Contact, systemic	Medium	Jassid, thrips	700 ml

[1] For many insecticides more than one formulation is registered; only one formulation is shown here.
[2] 1 acre = 0.404 ha.
[3] Chitin synthesis inhibitor.

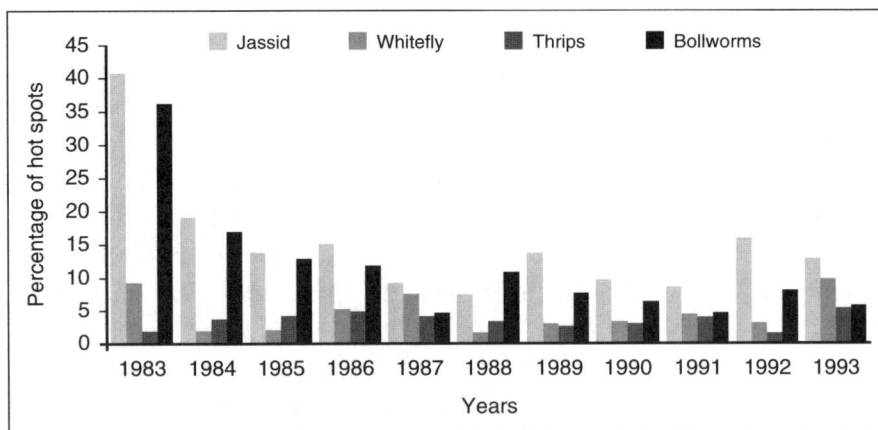

FIGURE 1
Comparative pest infestations in Punjab, 1983 to 1993 (mean of weekly averages for the season)

Table 15 shows that now nearly all farmers in Punjab spray their cotton, most applying three or four sprays, although increasingly six sprays are applied. Similar information is not available for Sind but it is believed that some 60 percent of the cotton area is sprayed two or three times.

Under the IPM system the number of insecticide applications can often be reduced if they are based on economic threshold levels (ETLs) established

TABLE 15
Percentage of cotton growers applying different numbers of sprays in Punjab

Year	No sprays	1 spray	2 sprays	3 sprays	4 sprays	5 sprays	6 sprays
1987	28.88	19.75	22.83	23.44	4.28	0.82	0.00
1988	26.15	25.68	20.80	20.35	0.80	0.80	0.00
1989	9.18	15.71	17.89	25.80	18.21	13.21	0.00
1990	6.76	2.60	8.92	23.24	38.75	19.73	0.00
1991	0.81	1.67	2.07	42.20	36.01	17.24	0.00
1992	0.00	0.49	7.14	26.37	38.39	25.03	2.04
1993	0.00	2.94	5.05	26.94	29.44	25.32	9.63

Source: Directorate of Pest Warning and Quality Control of Pesticides, Department of Agriculture, Government of the Punjab, Pakistan

through regular pest monitoring. The Punjab Government's Directorate of Pest Warning and Quality Control of Pesticides now uses pest monitoring and forecasting methods developed by the Central Cotton Research Institute at Multan. Investigations at this institute have indicated that American bollworm has developed resistance to cypermethrin and some organophosphate insecticides, and there is evidence that whitefly is also developing resistance to a number of insecticides, which would explain the recently reported difficulties in controlling these pests. The increasing importance of late-season aphid infestations, which reduce lint quality by honeydew excretion, is attributed to the indiscriminate use of insecticides.

The effect of pesticides on the health of farmers and labourers who come into contact with them is currently being investigated. No work on pesticide residues in cottonseed and seed cotton has been undertaken in Pakistan.

Pesticide application methods still leave room for considerable improvement. Large farmers use tractor-mounted boom sprayers while backpacks are favoured by small farmers. Sprayer quality is generally poor and equipment is poorly maintained and calibrated. The result is uneven distribution of pesticides on the crop, often leading to outbreaks of pests that are normally of secondary importance, such as spider mite, American bollworm and leafworm.

Legislative control

Federal. Under the provisions of the Plant Quarantine Act of 1976 the import of all plant material must be accompanied by a valid certificate of health (phytosanitary certificate) from the country of origin and permission to import (import permit) from the Pakistan government to prevent the entry of exotic plant pests and diseases. Special restrictions have been imposed on the import of potatoes, rubber, sugar cane, tobacco, citrus, coffee, bananas, coconut, groundnut, maize, tea, soil and rooted plants and cotton. The import of raw cotton produced in any part of the western hemisphere (North, South and Central America and the adjoining islands) is prohibited in order to exclude the entry of cotton boll weevil (*Anthonomus grandis* Boheman Coleoptera: Curculionidae). Cottonseed can be imported, in quantities not exceeding 1 pound (0.454 kg), by government research organizations only. Special permission for importing even this quantity is mandatory.

The Agricultural Pesticides Ordinance of 1971 and the Rules of 1973 regulate the import, manufacture, formulation, sale, distribution and use of pesticides. All pesticides are required to be registered by the government before being used in Pakistan. Products are registered after field trials for at least two crop seasons. During this period the spectrum of activity and toxicity to non-target organisms is taken into account. Pesticides are registered for use against specific pests only.

The Pesticides Ordinance of 1989 permitted the import of generic products such as monocrotophos, dimethoate methamidophos, cypermethrin and triazophos, without the necessity of each new supplier having to register the pesticide anew, provided that the original product had already been registered and that samples were analysed by the Department of Plant Protection before import. The purpose of the ordinance was to reduce the cost of pesticide imports, where products were out of patent, by obtaining them from the cheapest source commensurate with the maintenance of quality.

The Pesticides Ordinance was further liberalized in 1992 to allow the import of pesticides provided that products were registered in the country of manufacture. Such products may now be imported without going through the registration process. At the same time penalties for pesticide adulteration were increased, with higher fines or imprisonment for terms ranging from six months to four years.

Provincial. Under the West Pakistan Pest Control Ordinance of 1959 provincial governments are empowered to enforce certain cultural control practices and, in emergencies, to take appropriate action and charge the cost to farmers. The ordinance covers *Agrotis* sp., *Aphis gossypii, Earias insulana* and *E. vitella, Pectinophora gossypiella, Sphenoptera gossypii* Cotes (Coleoptera: Buprestidae) and *Syllepte derogata* (F.) (Lepidoptera: Pyralidae). Cotton stalks must be cleared from the field by 31 January in the Hyderabad and Sukkur divisions of Sind and by 15 February elsewhere in the country.

Cultural control

Crop rotation. Traditionally the main rotations in the cotton areas have been cotton-fallow and cotton-wheat and these have served to suppress many cotton pests. More recently other crops have been added to the rotation, including sunflower, soybean and maize, some of which act as alternate hosts for American bollworm, whitefly and other cotton pests. There was an outbreak of American bollworm in the 1990/91 season, when damage levels reached 10 percent. The outbreak was attributed to increased breeding by the bollworm on alternate hosts in the rotation. Okra is a prefered host of spotted bollworm, populations of which build up on adjacent okra fields before moving to cotton. Harvesting and the destruction of okra residues before cotton is sown reduce spotted bollworm attack and may save one or two early-season sprays.

Crop hygiene. As already noted pink bollworm is carried over from one season to another in left-over bolls on cotton stalks saved for fuel. Grazing stalks or burning them reduces carryover, indeed where stalk destruction has been carried out effectively pink bollworm may be eliminated completely. Ginnery waste can also be a source of pink bollworm carryover, but it is subject to predation by birds and high temperatures in April and May, which destroy diapause larvae. Ginnery waste is also sold as fuel to owners of brick kilns or to farmers as manure. These uses also destroy larvae so ginnery waste is not usually considered important in pink bollworm carryover.

Alternate host plants. The role of okra in the rotation as a source of infestation of spotted and spiny bollworm on cotton has already been noted. The two species differ in their host plant preferences; *E. vitella* favouring okra and kenaf (*Hibiscus cannabinus* L.) while *E. insulana* is commoner on ornamentals such as hollyhock *(Althaea rosea), Hibiscus rosa-sinensis, Abutilon indicum* and Turk's Cap *(Malvaviscus arboreum)*, all members of the Malvaceae. Up to seven generations may occur in the course of a year, with, in the case of *E. vitella,* the first generation following winter hibernation feeding on okra in the spring.

Sowing date. Selection of sowing dates for a particular variety of cotton is a matter of balancing a number of factors, including the need to ensure a sufficently long growing period and the desirability of avoiding pest attack. The main emergence of pink bollworm adults from diapause should occur before fruiting points are available for larvae to feed on. Very early sowing may attract infestations of thrips, jassid and spotted bollworm.

Line sowing. This is now the normal practice in Pakistan, although as recently as 1983 only 40 percent of the crop was line-sown, the remainder being broadcast. Line sowing enables a higher plant population to be obtained and facilitates weed control, which in turn reduces pest infestation.

Fertilizer. Overuse of nitrogen (N) on cotton results in lush growth which encourages American bollworm and aphid infestations and delays crop maturation, in turn allowing the buildup of pink bollworm populations. Late application of N can induce attack by aphid.

Irrigation. In Punjab irrigations applied after 15 October do not increase yield but delay maturation and enhance attack by pink bollworm.

Weed control. Where this is inadequate pest attack, particularly by sucking pests and leafworms, is more serious and the application of insecticides is less efficient than in clean fields, resulting in poorer control.

Manual control. This can be employed to control leafworm by hand-collection of larvae or egg masses, which can reduce populations by up to 80 percent.

Biological control

It is now recognized that natural enemies can exert control of many species of cotton pest but the sprays needed to control other pests destroy these natural enemies, resulting in the resurgence of key pests and the emergence of secondary pests. In the IPM system the strategy is to enhance the effectiveness of natural enemies by various methods, including delaying early sprayings, using selective or "soft" insecticides or using systemic insecticides that, when applied as seed dressings or as granules in the soil, do not affect beneficial species. Intercropping with crops such as sorghum or mung bean provides refuges for beneficials. Mass rearing and the release of parasitoids and predators have not so far been attempted on cotton in Pakistan.

Varietal resistance

Several morphological and physiological characters of the cotton plant confer resistance, to a greater or lesser degree, to one or more cotton pests. Some characters giving resistance to one pest, however, may increase susceptibility to another, and whether the character is used or not will depend, in part, on the relative importance of the two species and whether alternative control methods exist for one or other of them.

In cotton-breeding programmes in Pakistan a number of resistance characters are utilized. All new varieties have to have sufficient leaf hair to confer resistance to jassid. The nectariless character makes cotton less attractive to American and pink bollworm and some sucking pests, although it may also remove a source of food for some predators. Experiments at Multan with nectariless cottons have shown a reduction in pink bollworm infestations. The introduction of early-maturing and short-season varieties into commercial cotton cultivation since 1983 has reduced the significance of pink bollworm. Prior to this most commercial varieties were tall, monopodial and late-maturing.

The multi-adversity-resistance (MAR) genetic improvement system developed at Texas A&M University has proved of great benefit to Pakistan breeding programmes and, using MAR procedures, varieties have now been produced that are immune to bacterial blight, heat-tolerant and high-yielding. At the Central Cotton Research Institute (CCRI) at Multan breeding stocks with a variety of resistance characters, including high gossypol and glabrous, are maintained and are available to cotton breeders at other centres for hybridization and selection programmes.

Pheromones

The identification and synthesis of bollworm sex pheromones in recent years has opened up new possibilities for the operation of IPM programmes. Pheromones can be used in various ways, including for population monitoring, mass trapping and mating disruption.

The use of pheromones for monitoring has increased understanding of the activities of bollworms and may be a method of determining the timing of spray applications or mass releases of parasitoids. Figure 2 shows the average weekly trap catches of pink bollworm male moths at two sites in the Punjab between 1979 and 1993.

Mass trapping for pink bollworm control has produced erratic results in Pakistan. In 1978 four traps per acre (0.404 ha) appeared to reduce infestations for a time but immigration of mated females into the trapping area late in the season resulted in an increase in larval numbers.

Mating disruption is achieved by saturating an area with pheromone, released from point sources, to inhibit the location of unmated females by males. It offers a promising method of bollworm control and has been particularly successful with pink bollworm which, being a weak flier, is less

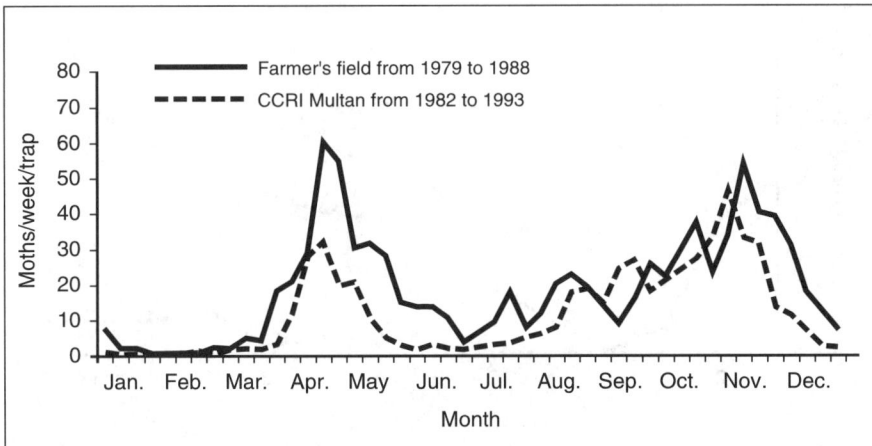

FIGURE 2
Pink bollworm male moth catches with pheromone-baited traps at CCRI, Multan, and in a farmer's field

prone to the problem of immigration by mated females from outside the disrupted area. Experiments on mating disruption using the pink bollworm sex pheromone gossyplure have been conducted for some years in Punjab.

Many different formulations have been evaluated, most recently PB Ropes, a slow-release formulation marketed by the Mitsubishi Company. This formulation gave season-long control of pink bollworm and similar results were obtained with Selebate, a formulation marketed by Agrisense. Figure 3 compares pink bollworm damage in pheromone- and insecticide-treated fields in the 1993/94 season.

INFRASTRUCTURAL SUPPORT FOR COTTON IPM
Federal
Pakistan Central Cotton Committee (PCCC). PCCC, under the Federal Ministry of Agriculture, is the principal organization responsible for research and development of cotton, both pre- and postharvest. It has its own multidisciplinary research institutes, the main ones being located at Multan in

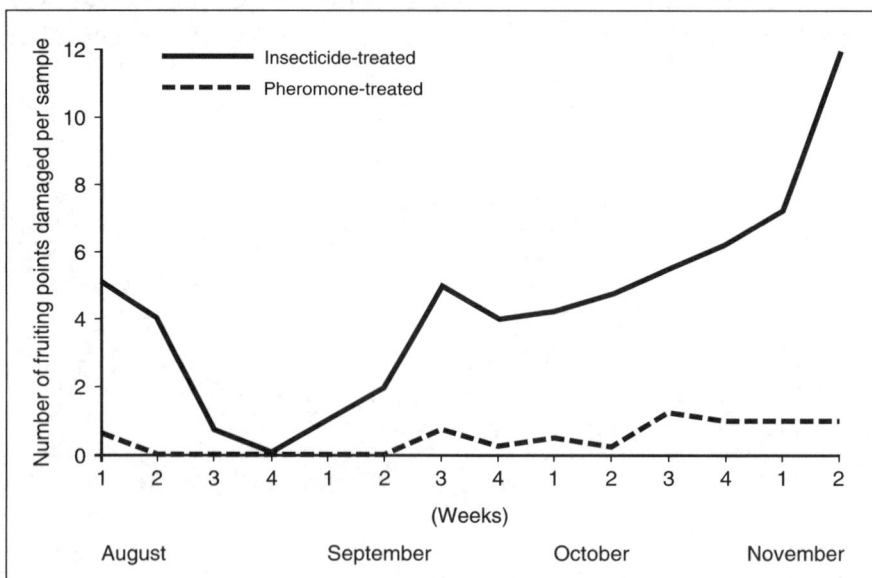

FIGURE 3
Comparative infestations of pink bollworm in insecticide- and pheromone-treated fields

Punjab and Sakrand in Sind. Both of these are chiefly concerned with preharvest research while another institute in Karachi covers postharvest cotton research.

The Multan and Sakrand institutes have been developing IPM programmes based on sucking pests and bollworms, natural enemies and crop development. Economic threshold levels have been established for the principal pests. Resistance to pesticides is currently being investigated as are new control techniques such as pheromones, while new varieties are screened for pest and disease resistance.

Plant Protection Department (PDD). PPD, part of the Federal Ministry of Agriculture, is responsible for pesticide registration and regulation under the various pesticide acts and ordinances. It is also responsible for the implementation of the 1976 Plant Quarantine Act to ensure that noxious pests, diseases and weeds are excluded from Pakistan.

PPD maintains a fleet of aircraft for aerial spraying. These are primarily used for locust control, but can also be used to control other major pest outbreaks on crops.

Atomic Energy Centre. The Atomic Energy Centre has two agricultural research institutes, one in Sind and the other in Punjab, involved in IPM. The main emphasis is on resistance in cotton to pests and diseases.

Pakistan Agricultural Research Council (PARC): International Institute for Biological Control. The institute is responsible for the collection, identification and rearing of parasitoids and predators for release in biological control programmes in various crops. It has identified the main beneficial species in cotton and has investigated the role they play in regulating populations of cotton pests.

Provincial

Each of the four provinces in Pakistan maintains agricultural research institutes, although only limited work is done on cotton IPM. Most of this is restricted to Sind and Punjab.

Ayub Agricultural Research Institute (AARI), Faisalabad. Located in Punjab, this is the oldest agricultural research institute in Pakistan and now includes a cotton institute together with five cotton research stations located in the main cotton-growing areas to carry out adaptive research under local conditions. At the main institute, which includes agronomy and sections for entomology, plant pathology, physiology and virology, both fundamental and applied research are undertaken.

Agricultural Research Institute, Tandojam. This is the main agricultural research institute for Sind. It has sections covering cotton breeding, agronomy, crop physiology, entomology and plant pathology. Most of the cotton pest research is of an applied nature and field oriented.

Plant protection services. The Punjab government maintains a Plant Protection Institute, primarily responsible for research on pesticides, and a Directorate of Pest Warning and Quality Control of Pesticides. This is located in Multan and employs 170 scouts who regularly monitor pest populations on various crops, enabling the directorate to issue weekly pest incidence reports to farmers and to advise them on appropriate control measures if pest populations reach economic threshold levels.

Extension services. In both Sind and Punjab there are well-organized extension services, headed by directors-general at provincial level, directors and deputy directors at regional and district levels.

The extension services work closely with the research institutes in the dissemination of information on cotton production technology and IPM. At the village level, extension workers provide farmers with advice, run training courses and grow demonstration plots.

Private sector
The private sector is mainly concerned with the sale of cottonseed, fertilizers, machinery and pesticides. Some of the larger companies, however, including Ciba and Hoechst, operate farmer training and extension programmes, mainly for cotton farmers, and provide pest scouting services. They have

done much to foster the IPM approach to cotton pest control. The Punjab Seed Corporation supplies cottonseed to farmers and also operates its own farm advisory service which provides farmers with assistance in plant protection matters.

ASSISTANCE TO COTTON IPM
Government assistance
Cotton Maximization Project 1974-1985. The cotton maximization project started as a pilot scheme on 7 000 acres (2 800 ha) of land in 1974. The aim was to increase cotton production through farmer training programmes and the provision of credit and other inputs. The target was to increase yields by 80, 160 and 240 kg seed cotton per acre (197, 395 and 590 kg per hectare, respectively) over the first three years of the project. Initial success led the government to extend the scheme to cover most of the cotton areas in Punjab and Sind. Extension workers involved in the scheme were trained at PCCC's Central Cotton Research Institute in Multan.

The scheme increased farmers' awareness of the new methods available to improve cotton production, including pest management technologies, making them receptive to further advances arising from the research programmes. By 1985 the government considered that the objectives of the project had been achieved and it was terminated. Yield increases in the project averaged 300 percent while the percentage of farmers controlling pests on their cotton with pesticides increased from 16 percent to almost 100 percent in the project areas compared with 60 percent in the non-project areas. Table 16 gives details of the scheme, including cost to benefit ratios which demonstrate the value of this project.

Foreign assistance
Integrated Pest Control Programme 1975-1980. This project (FG-Pa-2146-PK-ARS-46), supported by the United States Agency for International Development (USAID) and the United States Department of Agriculture using PL-480 funds, was executed by the Pakistan Agricultural Research Council, Islamabad. The aim was to investigate pest population dynamics in paddy rice, maize, sugar cane and cotton and to develop IPM programmes for these crops.

TABLE 16

Yield data and cost:benefit ratio in the Cotton Maximization Project areas

Year	No. of sites	Training		Area covered ('000 acres)[1]	Yields of seed cotton		Cost: benefit ratio
		No. of farmers	No. of extension workers		Project area	Non-project area	
1974/75	1			7	15.5	8.0	1:4.9
1977/78	6	12 723	206	60	9.2	6.8	1:4.9
1978/79	12	12 500	462	180	8.2	7.2	1:5.3
1979/80	12	12 625	469	180	12.7	9.7	1:4.5
1980/81	6	1 000	306	180	12.9	8.7	1:2.2
1981/82	6	1 095	296	181	11.7	8.2	1:4.4
1982/83	6	1 330	316	184	12.4	8.8	1:4.7
1983/84	3	1 575	215	91	6.7	5.7	1:3.8
1984/85	3	2 132	219	92	18.3	12.6	1:4.4
1985/86	3	1 974	253	92	22.1	18.4	1:6.2

[1] 1 acre = 0.404 ha

Strengthening the Central Cotton Research Institute, Multan. This project (Pak/73/026), funded by the United Nations Development Programme (UNDP) and implemented by FAO, commenced in 1974 and ran for a number of years. Research on cotton pest management was strengthened, scientists received overseas training and programmes were started on pest and natural enemy population dynamics and on the establishment of pest monitoring methods. Extension workers were trained in IPM technologies.

Integrated Pest Control of Cotton in Pakistan. This pilot project (CVR/76/7) was conducted in the 1976/77 season with funding from the United Kingdom Overseas Development Administration (ODA). The object was to prepare for a subsequent project to be funded by ODA (see "Cotton Pest Management Project, Pakistan" below) through pilot investigations in Multan and Tandojam on pesticide application methods for small-scale farmers.

Studies on the Sex Attractant of Pink Bollworm. This USAID project (PK-AR-182) ran from 1982 until 1984 using PL-480 funds. The use of pheromones to investigate the activity and control of pink bollworm was explored and it was concluded that pheromones could be used successfully for control.

Studies on the Development of MAR Cotton Lines in Pakistan. This USAID Project (PK-AR-184) ran from 1982 until 1984 using PL-480 funds. The project helped to establish the technologies for multi-adversity resistance (MAR) in cotton-breeding programmes in Pakistan, focusing on the breeding of new strains of cotton for bacterial blight resistance.

Strengthening the Research Capabilities of PCCC Institutes. This project (Pak.83/003) was financed by UNDP and executed by FAO. It was a follow-on from Pak/73/026 (see above) and extended the programme already initiated under the first project to cover, among other subjects, IPM research at PCCC's Sakrand Institute in Sind.

Cotton Pest Management Project, Pakistan. This project, a follow-up of CVR/76/7 (see above) was funded by ODA and executed jointly by ODA's Natural Resources Institute and PCCC. It was started in 1986/87 and combined surveys of farmers to determine the socio-economic context of cotton production, investigations aimed at improving pesticide application and studies on the use of pheromone to control pink and spotted bollworm. A summary of the main results of this project is given in the "Summary of results of the PCCC/ODA Cotton Pest Management Project" on p. 106.

Cotton Development Project. This project, started in 1987 with funding from the Asian Development Bank, is being executed by PCCC. The main objectives are:
  • improvement of cotton production through the supply of short-term credit to 20 percent of the farmers of the project area and support to the extension services of the provincial governments;
  • modernization of ginneries;
  • establishment of experimental ginneries;

- establishment of a Cotton Standards Institute;
- credit for the supply of agricultural inputs to the farmers in the project area.

Under the project many scientists have received training overseas in different aspects of cotton production technology. Pest monitoring systems are being established in Sind and North West Frontier Province.

KEY PESTS

The main pests of cotton in Pakistan are: pink bollworm *(Pectinophora gossypiella)*, spotted bollworm and spiny bollworm *(Earias vittella* and *E. insulana)*, jassid *(Amrasca devastans)*, whitefly *(Bemisia tabaci)* and aphid *(Aphis gossypii)*.

KEY PERSONNEL INVOLVED IN COTTON PEST MANAGEMENT IN PAKISTAN

Dr Zahoor Ahmad, Director, Central Cotton Research Institute, Old Shujabad Road, Multan, Punjab; specialist in integrated pest management.

Mr M. Rafique Attique, Senior Scientific Officer, Entomology Section, Central Cotton Research Institute, Multan, Punjab; specialist in integrated pest management.

Mr Mohammad Ali Chaudhry, Senior Scientific Officer, Entomology Section, Central Cotton Research Institute, Multan, Punjab; specialist in biological control.

Dr Mushtaq Ahmad, Entomologist, Central Cotton Research Institute, Multan, Punjab; specialist in pesticide resistance.

Dr Munir Ahmad Bhatti, Nuclear Institute for Agriculture and Biology, Jhang Road, Faisalabad, Punjab; specialist in varietal resistance.

Dr A.I. Mohyuddin, Director, PARC International Institute for Biological Control, Rawalpindi, Punjab; specialist in biological control.

Dr Ahmad Ali Baloch, Entomologist, Central Cotton Research Institute, Sakrand, Sind; specialist in integrated pest management.

Mr Mian Talib Hussain, Senior Scientific Officer, Plant Pathology Section, Central Cotton Research Institute, Multan, Punjab; specialist in multi-adversity resistance and cotton diseases.

Mr M. Islam Gill, Senior Scientific Officer, Agronomy Section, Central Cotton Research Institute, Multan, Punjab; specialist in weed control.

Dr M. Arshad Shakeel, Entomologist, CIBA Pakistan (Pvt.) Limited, Agriculture Division, 15 West Wharf, P.O.Box 100, Karachi, Sind; specialist in resistance monitoring.

Annex
SUMMARY OF THE RESULTS OF THE PCCC/ODA COTTON PEST MANAGEMENT PROJECT 1986-1990

The project received funds from the United Kingdom Overseas Development Administration (ODA) and was executed jointly by the Pakistan Central Cotton Committee (PCCC) and ODA's Natural Resources Institute (NRI). The main elements of the project included investigations into: the application and management of insecticides, with special reference to the needs of small farmers; general pest management, including the use of pheromones for bollworm control; and socio-economic surveys aimed at defining the target groups of farmers for cotton pest management research and extension and at identifying the constraints that affect the capacity of farmers to adopt new technologies.

Farmers growing cotton in Pakistan are not a homogeneous population, they range from resource-poor farmers on marginal land cultivating less than 1 ha to large-scale commercial farmers, growing perhaps hundreds of hectares of cotton and other crops, fully mechanized and often with other business interests.

The results of surveys involving interviews with farmers carried out over three years, suggested that cotton growers could be divided into four main groups based on their yield expectations and the land tenure system under which they farmed. There was a tendency for farmers in Punjab to be owner-occupiers while in Sind a landlord-tenant system, in which the farmers were in fact sharecroppers, was more common.

The first group includes these sharecroppers whose ability to increase production is likely to be constrained by the landlord-tenant relationship, by poor access to credit and other inputs and by shortages of irrigation water and problems with salinity.

Similar in many ways to the sharecroppers are the very small-scale, resource-poor farmers who, although they own their own land, are often located in marginal areas where problems of salinity and lack of water make the farming environment one of high risk. The yields of this group are usually less than 1 000 kg of seed cotton per hectare and the husbandry methods used are based on a low input-low output system. The use of pesticides is low and ownership of tubewells and tractors uncommon.

The majority of farmers probably fall into the third, medium-yield group (1 000 to 2 000 kg of seed cotton per hectare). Access to credit and other inputs is not such a serious constraint in this group, but lack of farm management skills and limited access to information on the use of inputs, especially pesticides, limits the efficiency with which resources are used and can result in high variable costs and low profitability in cotton growing.

Members of the small group of large-scale farmers obtaining yields of more than 2 000 kg of seed cotton per hectare (some exceed 3 000 kg per hectare) usually make efficient use of inputs, especially pesticides, and obtain higher gross margins than other groups.

These wide variations in the types of cotton grower and the constraints under which they operate have implications for the targeting of research and extension to meet their different needs and for the potential contributions the various groups can make to increasing national cotton production.

The sharecroppers and resource-poor farmers need special attention to be given to the removal of the constraints that prevent them from advancing their standard of living through the production of cash crops such as cotton. Many of these constraints are of a sociological or economic nature and are not therefore a matter for agricultural research. However, recommendations arising from research targeted at these two groups need to give growers an optimum mix of inputs that they can afford with their limited resources. Primary emphasis should be put on the use of pesticides. A rise in average yields of up to 40 percent over a five-year period should be achievable, increasing yields from an average of 500 to 800 kg of seed cotton per hectare.

The average yield of the medium-yield growers could be raised from 1 500 kg of seed cotton per hectare to 2 000 kg per hectare, on the basis of existing cotton production and pest management recommendations, through a programme of farm management training and extension. The scope for raising the already high yields of the fourth group of farmers is probably more limited in the medium term and the cost of research to make marginal improvements would be high. Over a five-year period a rise of 5 percent in the yields of this group is all that could reasonably be expected because a plateau of resource allocation and technical efficiency will have been reached.

Low- and medium-yield growers have the most potential for contributing to an overall growth in production provided research resources are switched to meet their particular needs. Such a policy would achieve a more equitable balance in the distribution of the benefits of research on cotton. However, for this to make an impact it is essential that the capacity of the extension services to reach small-scale farmers is strengthened.

The farmer surveys showed that, at all levels, pesticides were not being employed to the best effect and there is a need for improvement in product labelling and support. Cotton growers need to be empowered to make informed choices regarding pest control measures and need to be aware of the consequences of decisions, in terms of costs and benefits. To achieve this, researchers must provide a variety of options to meet local needs, taking into account the different resources available to different types of farmers.

Research priorities should include the development of practical pest scouting methods and a reassessment of the timing of control measures within the context of limited resources, where choices may have to be made by the farmer on the best use of a very small crop protection budget. Resistance to insecticides now poses a major threat to the sustainability of cotton production and needs greater research effort.

Another important component of the project was the evaluation of various methods of insecticide application on cotton. Trials were carried out over three seasons at Sakrand and Multan during which the performance of a range of lever-operated backpack sprayers, motorized mist blowers and ultra-low-volume sprayers were compared through physico-chemical monitoring and biological assessment. The object was primarily to meet the needs of small-scale growers but, at Multan, tractor-mounted boom and nozzle sprayers were compared with tractor-drawn mist blowers. These machines are in common use among large-scale farmers. Boom sprayers were found to give a more even, and therefore more efficient, cover of insecticide across the swath than mist blowers.

With the hand-held machines it was found that low-volume or ultra-low-volume (ulv) sprays applied by mist blowers and ulv sprayers respectively gave poorer droplet cover of the crop, as measured by physico-chemical methods, than high-volume sprays, although there was less difference in biological effectiveness. Investigations on different methods of application and different pesticides on whitefly populations indicated that control at ulv rates of application was poorer than that achieved at high volumes, while populations were much higher on plots sprayed with pyrethroids than on plots sprayed with organophosphorus insecticides or left untreated.

References

Ahmad, M. 1989. Identification of pest problems of pulses in Pakistan. *Pak. J. Sci. Res,.* 41: 25-31.

Ahmad, M. & Harwood, R.F. 1973. Studies on a whitefly transmitted yellow mosaic of urdbean (*Phaseolus mungo*). *Plant Dis. Rep.,* 57(9): 800-802.

Ahmad, M. & Harwood, R.F. 1973. Colour preference as a population indexing technique in the whitefly *Bemisia tabaci* Genn. (Aleyrodidae: Homoptera). *Pak. J. Agri. Sci.,* 10: 19-24.

Ahmad, Z. 1976. Source of infestation of pink bollworm. *Proceedings of Cotton Production Seminar,* organized by Esso Pakistan Fertilizer Limited, 20 to 30 April 1976, Sukkur, Pakistan.

Ahmad, Z. 1979. Pest management with special reference to pest scouting and forecast. *Proceedings of the Seminar on Cotton Production and Protection,* 16 to 18 January, 1979, Agriculture Department, Lahore, Pakistan.

Ahmad, Z. 1979. Incidence of major cotton pests and diseases in Pakistan with special reference to pest management. *Proceedings of International Consultation on Cotton Production with focus on Asian Region,* 17 to 21 November 1980, Manila, the Philippines.

Ahmad, Z. 1991. *Cultural control as part of Integrated Pest Management Programme in Cotton – Pakistan.* Paper presented in FAO-PCCC Regional Workshop on IPM in Cotton, 25 to 28 February 1991, Karachi, Pakistan. 40 pp.

Ahmad, Z. 1991. Integrated pest control in Pakistan. *Proceedings of the Conference on Integrated Pest Management in the Asia-Pacific Region,* 23 to 27 September 1991, Kuala Lumpur, Malaysia. CAB International and Asian Development Bank. 459 pp.

Ahmad, Z. 1993. *Status of cotton IPM in Pakistan.* Final Report of First Planning Meeting of Project on Integrated Pest Management in Cotton, 24 to 29 May 1993, Kuala Lumpur, Malaysia. CAB International.

Ahmad, Z. 1993. Control of pink bollworm with Gossyplure in Punjab, Pakistan. *Proceedings Working Group Meeting,* IOBC/WPR5 Bulletin, 16(10): 141-148.

Ahmad, Z. & Vaughan, M.A. 1980. Pest scouting on cotton. *Pak. Agric.,* 2(8&9): 44-47.

Ahmad, Z., Attique, M.R. & Khaliq, A. 1983. Preliminary observations on the efficacy of various pyrethroids and organophosphates as cotton insecticides. *Pak. Cottons,* 27(2): 11-14.

Ahmad, Z., Attique, M.R. & Rashid, A. 1985. An estimation of the loss in cotton yield in Pakistan attributable to the jassid. *Crop Prot.,* 5(2): 105-108.

Ahmad, Z., Attique, M.R. & Shakeel, M.A. 1983. Relationship of age and moisture content of cotton bolls to pink bollworm attack. *Pak. Cottons,* 27(3): 123-126.

Ahmad, Z., Halimee, M.A. & Naqvi, K.M. 1988. Harmful pests of cotton and their control. In: *Cotton in Pakistan,* PCCC and FAO. 280 pp.

Ali, M. 1970. A note on some of our problems of pest control in cotton. *Pak. Cottons,* 14(4): 295-297.

Ali, M. 1971. Cotton production in Pakistan. *Pak. Cottons,* 15(4): 136-156.

Ali, M., Ahmad, Z. & Attique, M.R. 1982. Effect of the removal of fruiting parts of four different varieties of cotton in Punjab. *Pak. Cottons,* 26(3): 123-130.

Ali, M., Ahmad, Z., Attique, M.R. & Waheed, T. 1988. Effect of some insecticides on growth, fruit development and yield of two cotton varieties in Punjab. *Pak. Cottons,* 32(1): 30-36.

Ali, M. & Attique, M.R. 1984. Determination of cotton plant ability to recover from bollworm damage. *Pak. Cottons,* 27(2): 113-116.

Ali, M. & Attique, M.R. 1986. Economical spraying schedule for cotton bollworm control and effect on yield of seed cotton. *Pak. Cottons,* 30(1): 1-6.

Ali, M. & Attique, M.R. 1987. Growth and development of cotton as affected by different groups of insecticides in Punjab, Pakistan. *Pak. Cottons,* 31(3): 215-222.

Ali, M. & Rashid, A. 1988. Efficacy of some insecticides against cotton pests. *Pak. Cottons,* 32(2): 63-65.

Arif, M.I. & Attique, M.R. 1990. Alternate hosts in carry-over of *Earias insulana*

(Boisd.) and *Earias vittella* (F.) (Lepidoptera: Noctuidae) in Punjab. *Pak. Cottons,* 34: 91-100.

Attique, M.R. 1985. Pheromones for the control of cotton pests in Pakistan. *Pak. Cottons,* 29(1): 1-6.

Attique, M.R. & Ahmad, Z. 1990. Investigation of *Thrips tabaci* Lind. as a cotton pest and developmental strategies for its control in Punjab. *Crop Prot.,* 9: 469-473.

Attique, M.R., Ghaffar, A. & Rafique, M. 1989. Response of two-spotted spider mites *Tetranychus urticae* Lutch. (Tetranychidae: Acarina) to different insecticides. *Pak. Cottons,* 33(2): 76-81.

Attique, M.R. & Khaliq, A. 1983. Effect of cypermethrin and chlorpyrifos alone, and their combinations, on cotton pests. *Pak. Cottons,* 27(2): 133-136.

Attique, M.R. & Malik, M.N.A. 1982. Some observations on early-season damage to terminal buds of cotton plants by *Earias* species. *Pak. Cottons,* 26(4): 165-168.

Attique, M.R. & Rashid, A. 1983. Efficacy of pyrethroid pesticides for the control of cotton pests. *Pak. J. Agri. Res.,* 4(1): 65-67.

Attique, M.R. & Rashid, A. 1983. Evaluation of different doses of pyrethroids for the control of cotton pests. *Pak. Cottons,* 27(4): 267-270.

Attique, M.R., Rashid, A., Shakeel, M.A. & Ali, M. 1984. Ultra-low-volume spraying for the control of cotton pests. *Pak. Cottons,* 28(1): 35-37.

Attique, M.R. & Shakeel, M.A. 1982. Comparison of ulv with conventional spraying on cotton in Pakistan. *Crop Prot.,* 2(2): 231-234.

Baloch A.A., Soomro, B.A. & Mallah, G.H. 1982. Evaluation of some cotton varieties with known genetic markers for their resistance/tolerance against sucking and bollworm complex. *Turk. Bit. Kor. Derg.,* 6(1): 3-14.

Chamberlain, D.J., Ahmad, Z., Attique, M.R. & Chaudhry, M.A. 1993. The influence of slow release PVC resin pheromone formulations on the mating behaviour and control of the cotton bollworm complex (Lepidoptera: Gelechiidae and Noctuidae) in Pakistan. *Bull. Entomol. Res.,* 335-343.

Chamberlain, D.J., McVeigh, L.J., Critchley, B.R., Hall, D.R. & Ahmad, Z. 1994. *Use of pheromones to control three species of bollworm in cotton in Pakistan.* Paper presented at World Cotton Research Conference I, 14 to 17 February 1994, Brisbane, Australia.

Cheema M.A., Muzaffar, N. & Ghani, M.A. 1980. Biology, host range and incidence of parasites of *Pectinophora gossypiella* (Saunders) in Pakistan. *Pak. Cottons,* 24(1): 37-73.

Cheema M.A., Muzaffar, N. & Ghani, M.A. 1980. Investigation of phenology, distribution, host range and evaluation of predators of *Pectinophora gossypiella* (Saunders) in Pakistan. *Pak. Cottons,* 24(2): 139-176.

Cork, A., Chamberlain, D.J., Beevor, P.S., Hall, D,R., Nesbitt, B.F., Campion, D.G. & Attique, M.R. 1988. Components of female sex pheromone of spotted bollworm; *Earias vittella* F. (Lepidoptera: Noctuidae), identification and field evaluation in Pakistan. *J. Chem. Ecol.,* 14(3): 929-945.

Flint, K.M., Balasubramanian, M., Campero, J., Strickland, G.R., Ahmad, Z., Barral, J., Barbosa, S. & Khall, A.F. 1979. Pink bollworm response of native males to ratios of Z.Z.- and Z.E.-isomers of gossyplure in several cotton growing areas of the world. *J. Econ. Entomol.,* 72: 758-762.

Ghulam H.M., Baloch, A.A. & Soomro, B.A. 1987. Study on the occurrence, incidence and relative abundance of different species of thrips on cotton crop at Sakrand, Pakistan. *Pak. Cottons,* 31(2): 153-159.

Gill, M.I. & Anwar, M. 1981. Weed control in cotton. *Agric. Pakistan,* 3(10): 18-20.

Gill, M.I., Anwar, M. & Muhammad, D. 1985. Weed control in cotton and its economics. *Pak. Cottons,* 29(4): 202-209.

Gill, M.I., Muhammad, D., Anwar, M., Younas, M. & Zaki, S.M. 1988. Testing of machinery for the incorporation of pre-emergence herbicide in cotton. *Pak. Cottons,* 32(4): 234-240.

Gill, M.I. & Saunders, J.H. 1984. Control weeds in cotton with Stomp-330. *Pak. Cottons,* 28(4): 297-300.

Hussain, T. 1982. Disease resistance in cotton. *In:* K.A. Siddiqui & A.M. Faruqui, eds. *New genetical approaches to crop improvement,* p. 547-551. Karachi, Pakistan, PIDC Printing Press (Pvt.) Limited.

Hussain, T. 1984. *Final research report on development of multi-adversity-resistant (MAR) cotton lines in Pakistan.* Project PK-AR-184. Submitted to PARC/USDA.

Hussain, T. 1984. Prevalence and distribution of *Xanthomonas malcampestris* pv. *malvacearum* races in Pakistan and their reaction to different cotton lines. *Trop. Pest Manag.,* 30(2): 159-162.

Hussain, T. & Ali, M. 1975. A review of cotton diseases of Pakistan. *Pak. Cottons,* 19(2): 71-86.

Hussain, T., Ali, M. & Yaqoob, M. 1981. Studies on seedling diseases and seed treatment of cotton. *Pak. Cottons,* 25(4): 197-207.

Hussain, T. & Mahmood, T. 1988. A note on leaf curl disease of cotton. *Pak. Cottons,* 32(4): 248-250.

Hussain, T. & Tanveer, M. 1981. Prevalence of seedling disease of cotton in Pakistan. *Pak. Cottons,* 25(1): 17-19.

Hussain, T., Tanveer, M., Saleem, A. & Yaqoob, M. 1975. Some studies on the boll rot of cotton in Punjab. I. Isolation and pathogenicity of the associated organisms. *Pak. Cottons,* 19(3&4): 181-188.

Hussain, T. & Yaqoob, M. 1976. Some studies on microflora and seed treatment of cotton. *Pak. Cottons,* 20(2): 115-123.

Hussain, T. & Yaqoob, M. 1976. Screening of cotton cultivars against bacterial blight of cotton. *Pak. Cottons,* 21(1): 51-56.

Mohyuddin A.I., Khan, A.G. & Goraya, A.A. 1989. Population dynamics of cotton whitefly *Bemisia tabaci* (Gennadius) (Homoptera: Aleyrodidae) and its natural enemies in Pakistan. *Pak. J. Zool.,* 21(3): 273-288.

Montemayor, M.B. 1986. How to apply doses of pesticides on cotton using small sprayers. *Agric. Pakistan,* 8(1): 34.

Saleem, A., Tanveer, M. & Hussain, T. 1975. Fusaria of Pakistan cotton. *Pak. Cottons,* 19(3&4): 194-196.

Tanveer, M. 1976. A review of nematodes associated with cotton plant with special reference to their prevalence in Pakistan. *Pak. Cottons,* 21(1): 57-58.

Tanveer, M. 1979. The survey of nematodes in cotton fields of Sind and Punjab. *Pak. Cottons,* 23(4-a): 395-396.

Tanveer, M. & Haque, E. 1975. Prevalence of nematodes in cotton fields of Pakistan. *Pak. Cottons,* 20(1): 15-24.

Tanveer, M. & Haque, E. 1979. Some studies on stunting of cotton in Pakistan. *Pak. Cottons,* 23(4-a) 325-334.

Waheed, T., Ahmad, Z., Ali, M. & Khaliq, A. 1983. Preliminary studies on the effect of some granular insecticides on sucking pests of cotton. *Pak. Cottons,* 27(3): 177.

Waheed, T., Ahmad, Z., Vaughan, M.A. & Attique, M.R. 1979. Comparative effect of synthetic pyrethroids and endosulphan on cotton pests and cotton crop in Pakistan. *Pak. Cottons,* 23(3&4): 239-243.

The Sudan

N. Sharaf Eldin

INTRODUCTION

Cotton is the most important cash crop in the Sudan and provides the main source of foreign currency, which in 1989 exceeded US$200 million. Most of the crop is exported as raw cotton, the local textile industry utilizing only about 20 percent. Long-staple *Gossypium barbadense* varieties and medium-staple *G. hirsutum* varieties predominate and are grown under irrigation in a season lasting from July to March. The Sennar dam, built in 1925 on the Blue Nile, is the source of water for the various flood, pump and gravity irrigation schemes. There is some rain-grown cotton, mainly short-staple *G. hirsutum* varieties, produced in a narrow belt running east to west across the centre of the country. Rainfall is confined to the period from June to August and normally ranges from 250 to 400 mm. In recent years, however, the rains have frequently failed leading to a decline in rain-grown cotton production.

COTTON PRODUCTION

Cotton is grown on about 4 percent of the country's total cultivated area of about 8 million ha. In the decade to 1993 production declined by nearly one-third. This decline was most marked in the last three seasons as land was switched to cereal crops under a government policy to meet shortfalls in food production in the rain-fed areas (Jack and Sharaf Eldin, 1994). Yields are declining or, at best, stagnating (Table 17). The main constraints on cotton production include the continual increase in production costs, especially the costs of plant protection, poor yields and the limited number of commercial varieties.

TABLE 17

Cotton area, total production and yields, 1981/82 to 1992/93

Season	Area (*'000 ha*)	Production (*'000 bales*)[1]	Lint yield (*kg per ha*)
1981/82	376	712	413
1982/83	407	944	505
1983/84	415	1 021	536
1984/85	373	933	545
1985/86	338	651	420
1986/87	360	748	454
1987/88	330	622	411
1988/89	325	637	427
1989/90	289	584	440
1990/91	203	379	406
1991/92	152	303	433
1992/93	160	310	424

[1] 1 bale = 218 kg.

CROP MANAGEMENT

The main commercial varieties derive from Barakat (*G. barbadense*), Shambat and Acala (*G. hirsutum*). The allocation of varieties to different areas is partly determined by their susceptibility to the main diseases, bacterial blight and fusarium wilt. Shambat and Acala were resistant to both diseases but bacterial blight resistance has broken down in these varieties during the past year. All three varieties are equally susceptible to the insect pest complex. Cottonseed is dressed with an insecticide-fungicide mixture to protect seedlings from flea beetle, bacterial blight and wilt.

Land for cotton is prepared by ploughing, clod crushing, levelling and ridging. Sowing is by hand. Fertilizer (urea of 46 percent nitrogen) is applied at the rate of 190 to 285 kg per hectare (80 to 120 kg per feddan where 1 feddan = 0.420 ha) by machine six weeks after sowing. The crop is irrigated at 14-day intervals until the end of January (short-staple cottons) or the end of March (long-staple cottons). Cotton is picked by hand, the harvest period extending from December to February for the short-staple varieties and from January to March for the long staples.

Competition for labour from other crops has resulted in weed problems in cotton, causing serious losses in yield. For a time herbicides provided a solution to the problem, but their increasing cost has led to reconsideration of other methods of weed control, including mechanical methods, used either alone or in conjunction with reduced use of herbicides.

Since the mid-1970s the agro-ecosystem in the irrigated cotton areas has both intensified and diversified, providing an environment that is increasingly favourable to the cotton pest complex; increased vegetable production, for example, gives a wide range of suitable alternative and alternate host plants for the cotton insects. In the Gezira region the crop rotation is now cotton-wheat-sorghum or groundnut-fallow.

PESTS
Insect pests
In the past the normal pattern of infestation by insects on July-sown cotton commenced in August with attacks by jassid, populations of which reached spray thresholds by mid-September. American bollworm first appeared in the crop in early September and reached spray thresholds by October. Jassid and bollworm comprised the so-called early pest complex. The late pest complex, comprising whitefly, which appeared in early November, and aphid, which appeared as temperatures dropped from mid-November, then persisted until the end of the cotton season.

More recently this division into early and late pest complexes has begun to break down. Whitefly infestations now begin early, at the same time as jassid appears, while aphid may appear as an early pest in July or August if rainfall causes a drop in temperature. The former strategy of applying two early sprays to control the early pest complex is no longer effective as all the sucking pests persist until the end of the season. The control strategy is now based on the season-long management of the whole pest complex.

Bemisia tabaci *(Gen.) (Hemiptera: Aleyrodidae).* Cotton whitefly is the most important pest of cotton in the Sudan, reducing both yield and quality through direct feeding and acting as a vector for leaf curl virus. The latter was a serious problem in the 1950s, but varietal resistance has now reduced its importance. Honeydew, produced by whitefly and aphid, contaminates lint and causes serious problems in ginning and spinning.

Helicoverpa armigera *(Hb.) (Lepidoptera: Noctuidae).* American bollworm first emerged as a pest of cotton in the Sudan in the 1960s, and since then it has established itself as second only to whitefly in economic importance. It not only attacks cotton, where a single larva may destroy a large number of fruiting points during the course of its development, but also infests vegetable crops and many other field crops in the rotation. Heavy early infestations can completely remove the bottom crop, which normally gives the best grade of lint.

Jacobiasca lybica *(de Berg.) (Hemiptera: Cicadellidae).* Cotton jassid was first recorded as a serious pest in the Sudan in the 1940s and was the first target for insecticide applications which were initiated in the 1945/46 season. Jassid is a sucking pest, both nymphs and adults feeding on cotton leaves. The injection of saliva into the vascular system, which then becomes blocked, causes the characteristic yellowing and then reddening of the peripheral areas of the leaves, known as "hopper burn" (Schmutterer, 1969).

Aphis gossypii *(Glov.) (Hemiptera: Aphididae).* Cotton aphid is a sucking pest that used to be considered mainly a late-season pest causing honeydew contamination of the lint, often accompanied by the buildup of sooty mould (Sharaf Eldin, 1977). Its importance declined in the 1970s, but increased again in the 1980s when it became a season-long pest, affecting both yield and quality. This increase was probably a consequence of the application of pyrethroid insecticides for bollworm control. Cotton aphid is not known to be a disease vector in the Sudan.

Podagrica *spp. (Coleoptera: Chrysomelidae).* Cotton flea beetle attacks seedlings causing characteristic round holes ("shot holes") in the leaves. Heavy attacks may destroy the entire stand, especially in drought years in the rain-fed areas. It is less serious in irrigated areas and where insecticidal seed dressings are routinely used.

Other insect pests. Schmutterer (1969) lists a number of other insects attacking cotton in the Sudan, including: *Diparopsis watersi* (Roths.) (Lepidoptera: Noctuidae) red bollworm, *Earias insulana* (Boisd.) (Lepidoptera: Noctuidae) spiny bollworm, *Pectinophora gossypiella* (Saund.) (Lepidoptera: Gelichiidae) pink bollworm, *Caliothrips* spp. (Thysanoptera: Thripidae) cotton thrips, *Dysdercus* spp. (Hemiptera: Pyrrhocoridae) cotton stainers and *Microtermes* spp. (Isoptera: Termitidae) termites.

The first four species are the most serious, sometimes causing yield losses of up to 40 percent (Balla, 1982), while the others are of only occasional significance.

Diseases

Bacterial blight. Bacterial blight, caused by *Xanthomonas campestris* pv. *malvacearum* (E.F. Smith) Dye, appeared early in the history of cotton production in the Sudan and still represents a threat. Its incidence and spread are most serious in years of high rainfall. Heavy, early attacks can result in complete defoliation and delay in fruiting point formation. Short-staple cotton varieties have generally been resistant to bacterial blight but the recent appearance of more virulent strains has, to a certain extent, caused a breakdown of this resistance.

Fusarium wilt. Fusarium wilt, caused by *Fusarium oxysporum* Schlecht f. sp. *vasinfectum* Atk. Sny. & Hans., is less serious than bacterial blight. It occurs in almost all the cotton-growing areas. Short-staple cottons are considered to be very resistant to the disease, but the long-staple varieties, such as Barakat 82, are susceptible. The variety Barakat 90, which was bred recently, is, however, highly resistant and is now grown in areas heavily infested with the disease (Balla, 1989).

Weeds

The most important weed species include *Cynodon dactylon* (L.) Pers. (Graminae), *Cyperus rotundus* L. (Cyperaceae) and *Ischaemum afrum* (J.F. Gmel.) Dandy (Graminae). Yield losses to weed competition may be as high as 30 percent. The costs of hand-weeding may represent up to 45 percent of total production costs (Hamdoun, 1978); herbicides are an alternative, but, as already noted, are expensive.

CONTROL MEASURES
Chemical control
The use of insecticides on cotton in the Sudan began in the 1945/46 season when one application of DDT was made in September to give season-long control of the cotton jassid in the Gezira Scheme (Snow and Taylor, 1952). The use of DDT for jassid control continued throughout the 1950s and led to the emergence of whitefly as a pest of economic importance. Control of both jassid and whitefly necessitated mixing DDT with an organophosphorus insecticide, such as dimethoate or parathion. This practice continued for some years until American bollworm emerged as a major pest in the 1963/64 season. After that the number of spray applications started increasing as it became necessary to control all three pests and, later, the cotton aphid as well (Lazarevic, 1964). The number of spray applications now varies from season to season, depending on the intensity of the attack by the four main pests. Applications peaked at 8.5 sprays in the 1985/86 season and declined thereafter to an average of 3.02 sprays in 1993/94. The cost of crop protection as a percentage of total production costs has thus varied over time (Table 18).

Insecticides. Insecticides imported into the Sudan must be tested by technical committees before being approved for commercial use. In the 1993/94 season the total cost of all pesticides used in the Sudan amounted to some US$13 million. Table 19 shows the principal insecticides and insecticide mixtures currently in use. Because of the need to control the whole pest complex the use of mixtures is very common.

Herbicides. Until recently these were used on the entire cotton area. Control of both broad-leaved and monocotyledonous weeds was obtained through the use of mixtures, the most commonly used are listed in Table 19.

TABLE 18

Average number of sprays, cost of one spray and cost of crop protection as a percentage of total production costs in the Gezira Scheme, the Sudan

Season	No. of sprays	Cost of one spray *(Sudanese £)*[1]	Protection as % of total production costs
1981/82	6.78	8.26	20.2
1982/83	5.11	14.29	24.2
1983/84	5.46	21.63	27.7
1984/85	4.15	22.90	21.4
1985/86	8.50	30.48	33.3
1986/87	5.22	44.04	27.5
1987/88	5.67	46.21	24.7
1988/89	5.27	56.17	19.3
1989/90	4.40	69.80	15.3
1990/91	3.70	90.70	10.0
1991/92	4.75	273.05	18.8
1992/93	4.93	1 601.19	42.0
1993/94	3.02	-	-

[1] Sudanese £100 = US$0.17.

Seed dressings. These are applied to give protection from attack by bacterial blight, fusarium wilt and flea beetle. The main seed dressings that have been used are as follows: phenylmercury acetate + gamma (HCH 50: 100 g per kg) at rates of 1.0 kg per 150 kg seed; bronopol + gamma (HCH 80: 200 g per kg) at rates of 1.5 kg per 150 kg seed; phenylmercury acetate + heptachlor at rates of 1.0 kg per 150 kg seed (this last seed dressing is being phased out of use).

Insecticide application. Tractor-mounted sprayers were used when spraying began in the 1945/46 season, but as the area under cotton increased this method of application became impractical and from 1949 it was gradually replaced by aerial application. By the 1960s all cotton was aerially sprayed.

TABLE 19
Main insecticides and herbicides used in the Sudan

Insecticide and formulation	Application rate per feddan[1]
Insecticides	
Cypermethrin + chlorpyrifos 1.4:2.4 ul	1 500 ml
Ekatin combi[2] + 11:4.5:35 ul	1 000 ml
Deltamethrin + dimethoate 7:18 ul	1 000 ml
Cypermethrin + profenofos 2:20 ul	750 ml
Fenvalerate + dimethoate 5:18 ul	1 000 ml
Dicrotophos 50 ec	420 ml
Profenofos 400 ec	700 ml
Endosulphan 50 ec	800 ml
Chlorfenvinphos 50 ec	336 ml
Amitraz 20 ec	1 000 ml
Fenpropathrin 50 ec	750 ml
Methomyl 90 wp	0.23 kg
Demeton-s-methyl 25 ec	900 ml
Diafenthiuron 50 ec	252 ml
Lambda-cyhalothrin 5 ec	100 ml
Herbicides	
Pendimethalin 50 ec + diuron 80 wp	1 200 ml + 0.25 kg
Oxadiazon + diuron 50 sc	1 000 ml
Norflurazon + diuron 80 + 80 wp	0.5 + 0.25 kg
Dipropetryn + metolachlor 1.5:1.0 (Cotodon 400 ec)	2 500 ml
Fluometuron + metolachlor 1:1 (Cotoran multi 50 wp)	1.25 kg
Oxyfluorfen 24 ec + diuron 80 wp	500 ml + 0.25 kg

[1] 1 feddan = 0.420 ha.
[2] Thiometon plus contact/stomach acting insecticides.

Two methods are now used in aerial spraying – conventional low-volume (clv) and ultra-low-volume (ulv). In conventional low-volume applications sprays are applied at the rate of 21.6 litres per hectare (2 gallons per feddan), the comparatively high volume of liquid giving the good cover needed to control late-season attacks of whitefly and aphid. Ultra-low-volume spraying was introduced in the 1970/71 season and has been found particularly effective for early-season control of jassid and American bollworm, which both feed on the upper part of the plant. Some 40 percent of spray applications are now ulv. The method allows a larger area of cotton to be sprayed for each aircraft sortie because spray volumes are low in the absence of water, so the costs of application are lower too, in comparison with clv application.

Pest populations are monitored and sprays applied when economic threshold levels (ETLs) are reached. The ETLs for the main pests are currently:
- jassid, 100 nymphs per 100 leaves;
- American bollworm, 30 eggs and/or 10 larvae per 100 plants (three eggs are equivalent to one larva);
- whitefly, 600 adults per 100 leaves;
- aphid, 40 aphid-infested plants per 100 plants.

Insecticides are purchased and stocked on the assumption that normally two ulv and three clv sprays will be required each season, but the actual number of sprays applied depends on population levels as determined by pest monitoring.

Legislative measures
Although chemical control is easy to implement and normally gives rapid and positive results, it can cause environmental and pest management problems. Therefore non-chemical methods of control have been investigated, and many adopted, some with legal backing (Sharaf Eldin, Saleem and Omer, 1978).

The storage of seed cotton, untreated seed and cotton stalks is prohibited by law. Quarantine laws prohibit the import of noxious pests.

Cultural control
Crop sanitation. In 1935 measures to prevent the carryover from one season to the next of pink bollworm and bacterial blight were introduced and legally enforced throughout cotton areas. The measures entailed pulling up and burning spent cotton plants, together with leaves and other waste swept up from the fields, in order to destroy larvae and disease spores. These practices have been less rigidly enforced of late as pink bollworm is no longer a serious pest and there is resistance to bacterial blight in some cultivars.

Crop rotation. This is particularly effective in reducing the impact of soil pests and soil-borne diseases and is now mandatory in all the cotton-producing areas. The type of rotation practised is determined by local needs and the availability of the necessary inputs.

Resistant varieties. It was shown early on that leaf hair conferred resistance to jassid in cotton (Knight, 1952). However, it was also found that this character led to greater susceptibility to whitefly (Mound, 1965). Both pests are equally important and for a long time there seemed little prospect of achieving resistance to both within the same cotton variety. Interest in the problem was recently revived and a programme to combine characters for hair density and hair length was initiated, leading to the release of two promising new cultivars, Acala 82 (8A) and Acala 82 (8B), which are expected to alleviate the jassid problem (Jack, 1992).

All the *G. hirsutum* varieties are resistant to fusarium wilt and Acala and Shambat B are grown in areas known to be wilt-infested. The long staple Barakat 82 is highly susceptible and so can be grown only in areas known to be free of the disease. Recently Barakat 90, a wilt-resistant variety, has been bred, and it will help to meet the continuing demand for long-staple cottons for export.

Bacterial blight resistance, bred into the short-staple varieties, was thought to have solved the problem of this disease, but recently a very high incidence of the disease on all cotton varieties is thought to have been caused by the evolution of more aggressive strains of blight (Omer and Kheiralla, 1978).

Biological control
Although preliminary investigations were initiated by Gameel (1968;1971) on whitefly parasitoids and predators, there was little serious research into their effectiveness as control agents. In 1979 an integrated pest management (IPM) project, financed by the Government of the Netherlands and executed by FAO, commenced. The project focused on natural and biological control of whitefly and American bollworm. In the period 1988 to 1990, *Trichogramma pretiosum* Riley (Hymenoptera: Trichogrammatidae), an egg parasitoid of American bollworm, was introduced into the Sudan and

released into cotton at five sites. Munir *et al.*, (1992) reported that the parasitoid became well-established and that parasitism rates ranging from 19 to 68 percent were found. There was also a single release of a whitefly parasitoid, *Encarsia* sp. in 1988, but there was no follow-up to that release.

CONCLUSIONS

The control of the cotton pest complex is based on a combination of chemical and non-chemical methods. Insecticides are applied only when economic threshold levels are reached. In the 1993/94 season, these were almost doubled as a result of the findings of the IPM project (Abdelrahman, Bannaga and Dein, 1992). This would appear to be the main reason for reducing the number of applications by an average of approximately 1.9 sprays (Table 18). It is expected that these higher threshold levels will be maintained and perhaps raised even further. Reducing the number of spray applications allows for important savings to be made in pesticide purchases and stocking levels.

The sole use of DDT for jassid control over a period of many years is thought to have been responsible for the emergence of whitefly as a major pest, either through the eradication of natural enemies (Bindra and Abdelrahman, 1983) or because of a change in the whitefly's physiology leading to greater fecundity (Hassan, 1982). For some years now it has been the policy to alternate different pesticide groups in the application sequence, in order to try to minimize such problems and delay the onset of pesticide resistance. As a result, chemical control is still giving good results, although organophosphorus resistance is known to exist in the whitefly population (Dittrich, Hassan and Ernst, 1985).

To solve the problem of resistance, which can limit the effectiveness of a pesticide to between five and seven years, an intensive programme of screening new compounds is being undertaken by theAgricultural Research Corporation. Priority is now given to seed dressings and granular formulations which are considered softer on beneficial species than foliar applied insecticides. There are additional benefits in the reduced impacts on human health, non-target species and the environment generally. A new strategy, based more on the use of non-foliar methods of application, could lead to a reduction in aerial spraying which is costly and, in some respects, inefficient.

The possibility of growing cotton without the use of insecticides was investigated under the IPM project. It was found that, while losses to jassid and American bollworm were very heavy, whitefly and aphid were controlled by their natural enemies. This was in contrast to treatment with insecticide where these latter two pests reached economic threshold levels (Eveleens and Abdelrahman, 1980). These findings emphasize the importance of delaying the first spray for as long as possible to allow natural enemy populations to build up early in the season. The problem of jassid and American bollworm attack remains to be solved in such a way that aphid and whitefly infestations are not encouraged.

The problem is being tackled in three ways. The long-term approach is through breeding for resistance. Success has been achieved, as already noted, with jassid resistance; work on resistance to bollworms, with encouraging preliminary results, was begun, but not followed up (Khalifa, 1978).

The second approach, using biological control against American bollworm, was initiated under the IPM project, but needs follow-up if success is to be achieved. So far little serious work on natural control of jassid has been undertaken. It should be emphasized that natural control is the core of the IPM approach and much more work is needed to achieve an understanding of the factors involved.

The third approach is the search for "soft" insecticides which have minimal impact on the beneficial species that control aphid and whitefly, but give control of jassid and American bollworm. A larvicide specific to the latter is a possibility, but studies on jassid control have only recently been initiated.

In conclusion, therefore, it is suggested that the way forward for the development of IPM for cotton in the Sudan is to concentrate insecticide work on the use of seed dressings and granular formulations applied to the soil to control early sucking pests without harming beneficial species. These, if supplemented by augmentative releases of natural enemies, might be sufficient to inhibit the development of American bollworm populations. Foliar sprays could then be reduced to a minimum, creating an environment where natural enemies could be the main control agent for whitefly and aphid later in the season.

KEY PESTS

The main pests of cotton in the Sudan are: whitefly *(Bemisia tabaci)*, American bollworm *(Helicoverpa armigera)*, jassid *(Jacobiasca lybica)*, aphid *(Aphis gossypii)*, bacterial blight *(Xanthomonas campestris)* and fusarium wilt *(Fusarium oxysporum)*.

References

Abdelrahman, A.A., Bannaga, A. & Dein, Y. 1992. *Proceedings of the 58th Pests and Diseases Committee*, September 1992.

Balla, A.N. 1969. Some aspects of the ecology, biology and control of the American bollworm in the Sudan Gezira. *In:* M.A. Siddig & L.C. Hughes, eds. *Cotton growth in the Gezira environment*, p. 281-292. Wad Medani, the Sudan, Agricultural Research Council.

Balla, A.N. 1982. Progress in research and development for *Heliothis* management in the Sudan. *In:* W. Reed & V. Kumble, eds. *Proceedings of the International Workshop on* Heliothis *Management,* 15 to 20 November 1981, ICRISAT Center, Patancheru, AP, India.

Balla, A.N. 1989. Breeding cotton varieties for resistance to pests. *Annual Report of the Agricultural Reseach Corporation of the Sudan 1988-89.*

Bindra, O.S. & Abdelrahman A.A. 1983. *Integrated pest control of cotton in the Sudan.* Field Document No. 11 GCP/SUD/025/NET, FAO. Rome, Italy, FAO.

Dittrich, V., Hassan, S.O. & Ernst, G.H. 1985. Sudanese cotton and the whitefly: a case-study of the emergence of a new primary pest. *Crop Prot.,* 4: 161-176.

Eveleens, K.G. & Abdelrahman, A.A. 1980. *The cotton whitefly problem: can the tide be turned?* Pests and Diseases Committee.

Gameel, O.I. 1968. Studies on whitefly parasites, *Encarsia lutea* (Masi) and *Eretmocerus mundus* Mercet (Hymenoptera: Aphelinidae). *Rev. Zool. Bot. Afr.,* 79: 65-77.

Gameel, O.I. 1971. The whitefly eggs and first larval stages as a prey for certain phytoseiid mites. *Rev. Zool. Bot. Afr.,* 84: 79-82.

Hamdoun, A. 1978. *Prospects of an integrated control policy against terrestrial weeds.* Symposium on Crop Pest Management, February 1978, Khartoum, the Sudan.

Hassan, S.O. 1982. Investigations of DDT residues on cotton leaves on fertility of the cotton whitefly, *Bemisia tabaci*. Reading University, Reading, UK. (M.Sc. thesis)

Jack, I. 1992. Cotton cultivars resistant to Jassid. Varieties Release Committee.

Jack, I. & Sharaf Eldin, N. 1994. Future of cotton in the Sudan. Conference, March 1994, Khartoum, the Sudan.

Khalifa, H. 1978. Breeding cotton for insect pest resistance in the Sudan. Symposium on Crop Pest Management, February 1978, Khartoum, the Sudan.

Knight, R.L. 1952. The genetics of jassid resistance in cotton. The genes HI and HZ. *J. Genet.*, 51: 47-66.

Lazarevic, B.M. 1964. *Annual Report of the Agricultural Research Division, the Sudan, 1963/64.*

Mound, L.A. 1965. Effect of leaf hair on cotton whitefly populations in the Sudan Gezira. *Cott. Gr. Rev.*, 42: 33-40.

Munir, B., Abdelrahman, A.A., Mohammed, A.H. & Stam, P.A. 1992 Introduction of *Trichogramma pretiosum* Riley against *Heliothis armigera* (Hb.) in the Sudan. *Proc. of the 3rd International Conference on Plant Protection in the Tropics*, 5: 70-73. 20 to 23 March 1990, Genting Highlands, Malaya, Malaysian Plant Protection Society.

Omer, M.E.H. & Kheiralla, A.I. 1978. Bacterial blight of cotton (*Xanthomonas malvacearum*). Symposium on Crop Pest Management, February 1978, Khartoum, the Sudan.

Schmutterer, H. 1969. *Pests of crops in Northeast and Central Africa with particular reference to the Sudan.* Stuttgart, Germany, Gustav Fischer Verlag. 296 pp.

Sharaf Eldin, N. 1977. Late cotton insect pest control. *Annual Report of the Agricultural Research Corporation of the Sudan 1976-77.*

Sharaf Eldin, N., Saleem, B. & Omer, M.E.H. 1978. Non-chemical pest control practice. Symposium on Crop Pest Management, February 1978, Khartoum, the Sudan.

Snow, O.W. & Taylor, J. 1952. The large-scale control of the cotton jassid in the Gezira and White Nile areas of the Sudan. *Bull. Entomol. Res.*, 43: 479-502.

Syrian Arab Republic

Farid Khouri

INTRODUCTION

It is not known when cotton was first introduced into the Syrian Arab Republic. Until 1923 the area under cotton was less than 800 ha but by 1937 it had increased to 30 491 ha. Cotton production increased dramatically in the years following the Second World War. By 1951 the cotton area was 217 000 ha and it reached a peak of 302 000 ha in 1962. Over the past decade the area has varied from 155 000 to 170 000 ha. Yields have also increased from a 1945 average of 245 kg of lint per hectare to an average of 1 140 kg per hectare in 1993.

Production system

The government decides each year how much cotton will be grown and announces its decision at least four months before the cotton season begins. The planned area is allocated at farm level by the extension services. Thus, although cotton growing is largely in the private sector it is tightly controlled by the government. Individual farmers need a licence to grow cotton; this is obtained from the Department of Agriculture. The licence enables farmers to obtain credit from the Agricultural Cooperative Bank for the purchase of inputs including seed, fertilizer, pesticides and sacks.

During the growing season the extension services are responsible for advising farmers, for pest monitoring and decision-making on control measures and for giving regular estimates of yield potential to the government. Farmers are obliged to follow the advice of extension officers and may be fined if they fail to do so. Farmers may also lose their licence to grow cotton if they do not comply with its conditions.

The overall coordination of the cotton sector is achieved through an annual cotton congress which is attended by the Minister for Agriculture and

Agrarian Reform, together with some 100 to 120 officials. They review the achievements of the past season, including research results, and formulate policy for the future. The Cotton Bureau produces a bulletin containing recommendations for the coming season.

The Cotton Bureau is a directorate of the Ministry of Agriculture and Agrarian Reform (MAAR) and is responsible for the supervision of the cotton sector. It is not directly represented in the provinces, but is professionally responsible for cotton specialists in each provincial department of agriculture. The bureau operates a system of central committees of senior officers who, during the cotton season, spend two weeks each month in the cotton areas. The Cotton Bureau is also responsible for cotton research and the control of cotton exports. It produces extension material, including leaflets for farmers, television programmes and videos.

Since 1965 cotton production has been very successful. The main factors contributing to this success include:

- Cotton growers must be licensed by the Ministry of Agriculture and Agrarian Reform.
- The Cotton Marketing Organization purchases all seed cotton from farmers at a price for the basic quality grade that is announced before the beginning of each season. In 1991 the price was Syrian Piasters 2 000 per kilogram.
- Cotton growers can now obtain credit for the purchase of inputs, including seed, fertilizer, pesticides and sacks, from the Agricultural Cooperative Bank of Syria.
- The sowing period for cotton is controlled, with the last date permitted for sowing specified for each region. This is usually in mid-May.
- Extension units, which are responsible for providing advice on pest control and for yield estimations, cover well-defined, limited areas, thus ensuring their effectiveness.
- The mechanization of cotton production is now promoted by the newly established General Organization for Agricultural Mechanization in cooperation with the Cotton Bureau. The aim is to reduce production costs.

Cotton production

Table 20 shows the cotton area, production and yield for the seasons 1980/ 81 to 1993/93. For 1994/95 it was planned that 200 000 ha be grown, producing 230 000 tonnes of lint at an average yield of 1 150 kg per hectare.

Most cotton is grown in one of the three main cotton-growing regions of Syria:

- central-northern region (Aleppo and Hama provinces): 25 percent of the total cotton-growing area; growing period of 190 days; and average yields of 1 050 to 1 150 kg of lint per hectare;
- eastern region (Rakka and Deir Ezzore provinces): 45 percent of the total cotton-growing area; growing period of 185 days; and average yields of 850 to 950 kg of lint per hectare;
- northeastern region (El-Hassaka Province): 30 percent of the total cotton-growing area; average yields of 1 100 to 1 200 kg of lint per hectare;

Table 21 gives meteorological data for these regions during the cotton-growing season.

Socio-economic factors

Cotton is second only to crude oil as an earner of foreign exchange for Syria and its contribution to the gross national product (GNP) is exceeded only by oil and wheat. Currently 75 percent of the crop is exported. The Syrian population numbers 11 million, 1.8 million of whom are dependent for their livelihood on the cotton sector, including production, processing, manufacture and input supply. Cotton occupies one-third of the total irrigated area in Syria.

Cotton growth and development

Between sowing and final harvest the growth and development of the cotton plant can be conveniently divided into three periods: plant establishment is the period between sowing and the appearance of the first square; fruit formation goes up to the time of first boll split; and maturation goes up to final boll opening.

For the variety Aleppo 40, grown under normal conditions in Aleppo

TABLE 20

Cotton area, production and yield in the Syrian Arab Republic

Season	Area (ha)	Production (tonnes)	Lint yield (kg per ha)
1980/81	138 810	117 846	849
1981/82	143 433	129 502	903
1982/83	158 779	157 995	995
1983/84	172 559	194 000	1 124
1984/85	178 450	152 785	856
1985/86	170 200	161 745	950
1986/87	144 286	125 931	873
1987/88	128 688	96 489	750
1988/89	171 000	114 072	667
1989/90	158 000	128 237	812
1990/91	145 000	140 000	965
1991/92	170 000	194 000	1 143
1992/93	212 000	241 000	1 137
1993/94	196 000	223 000	1 140

province, the timing of the main events in the life of the cotton plant is as follows:

- germination – 12 to 15 days;
- first true leaf – ten days after germination;
- second true leaf – three days later;
- germination to first square – 42 to 45 days;
- square to white flower – 21 to 23 days;
- flower to boll split – 55 to 60 days (early to mid-season; 65 to 75 days late season);
- interval between flowers on same sympodium – six days;
- interval between flowers on successive sympodia – three days;
- time between first and last boll split – 70 to 75 days.

The same variety grown in the eastern region completes its development five to seven days earlier. The variety Aleppo 33/1 is similar to Aleppo 40

TABLE 21

Cotton season meteorological data for the main cotton-growing regions of the Syrian Arab Republic (monthly means)

Region	Month						
	April	May	June	July	Aug.	Sept	Oct.
Central-northern region (maximum latitude 36°N)							
Rainfall *(mm)*	25	15	2	0	0	3	14
Maximum screen temp. *(°C)*	23	29	34	36	37	34	28
Minimum screen temp. *(°C)*	10	14	18	20	20	18	13
Relative humidity *(%)*	60	55	50	52	55	55	60
Eastern region (maximum latitude 36°N)							
Rainfall *(mm)*	22	9	0	0	0	0	8
Maximum screen temp. *(°C)*	26	32	37	40	40	35	29
Minimum screen temp. *(°C)*	12	17	22	25	25	20	14
Relative humidity *(%)*	50	45	30	30	40	40	50
Northeastern region (maximum latitude 37°N)							
Rainfall *(mm)*	45	19	1	0	0	1	12
Maximum screen temp. *(°C)*	24	31	37	40	40	35	29
Minimum screen temp. *(°C)*	9	14	19	22	21	16	11
Relative humidity *(%)*	55	50	40	35	35	40	55

in its behaviour in all regions while the varieties Rakka 5 and Deir Ezzore 22 are seven to ten days earlier.

CULTURAL PRACTICES
Land preparation
Soils in the central-northern and northeastern regions are of the gromosol type (dark-red, brown, dark-brown, black montmorillonite clays), while in the eastern region the soils range from alluvial, grey, sandy loams to clays.

In the central-northern region field size is small, from 2 to 5 ha, and cotton is grown in a two-year rotation with wheat, sugarbeet, potatoes, legumes or vegetables, creating a very diverse agro-ecosystem. In the other regions field size is larger, at 10 to 25 ha, and the two-year rotation is based on cotton-wheat. Typically wheat is harvested in June, the land ploughed twice in opposite directions and then left to be weathered until the third ploughing which takes place in the following February or March. Following ploughing a seed bed is prepared and the field is levelled.

Cotton sowing is mainly by hand, although some single and multirow drills are in use. Sowing takes place from the beginning of April to mid-May. Inter-row spacing is 65 to 75 cm, and intrarow spacing is 18 to 25 cm. Between four and ten seeds are placed in each hole, giving a seed rate of 50 to 80 kg per hectare. On some 25 percent of the cotton area the seed is still broadcast, at 100 to 150 kg per hectare. Weeds are controlled both manually and with pre-emergence herbicides.

Irrigation
Irrigation water is obtained from wells, rivers and lakes and reaches fields via earth channels. Basin or flood irrigation is practised on about 60 percent of fields, most of the remainder being furrow irrigated. Overhead sprinkler irrigation is practised on a limited area. Between April and September cotton is irrigated between eight and ten times, giving a seasonal average consumption of 7 000 to 9 000 m^3 of water per hectare.

Fertilizer
The Soils Directorate, in cooperation with the Cotton Bureau, is responsible for fertilizer recommendations for cotton. Farmers apply between 75 and 95 kg of phosphatic fertilizer per hectare and 150 to 190 kg of nitrogen per hectare. Applications of nitrogen are usually split, with one-third applied at sowing and two-thirds at the start of flowering. This practice is designed to avoid rank growth and to meet the heavy nutritional demands of the plant during the reproductive period. Phosphatic fertilizer is applied at the time of sowing. Fertilizers are either broadcast by hand or applied by machine. Farmers obtain fertilizers from the Agricultural Cooperative Bank, in quantities determined by the conditions of their licences; this ensures application at the recommended rates.

VARIETIES
Medium-staple *Gossypium hirsutum* varieties are grown in Syria. In each region varieties adapted to local conditions, including varieties with resistance

TABLE 22
Performance of Syrian commercial cotton varieties (means of three years)

Variety	Yield *(kg seed cotton per ha)*			Wilt resistance factor	Gin (%)	Earliness (days from first flower to first pick)	Fibre properties			
	Mean of six sites	Mean of three sites (no verticillium wilt)	Mean of three sites (verticillum-wilt infested)				Length *(inch)*	Pressley	Strength *(g/tex)*	Micronaire
Aleppo 40	4 070	4 230	3 890	1.61	39.13	72	1.133	8.82	21.44	4.67
Rakka 5	4 330	4 110	4 490	0.93	38.25	76	1.118	8.68	22.57	4.68
Aleppo 33/1	3 620	3 380	3 910	1.38	38.94	68	1.206	9.38	25.14	4.36
Deir Ezzore 22	3 710	4 220	3 070	1.90	41.20	76	1.172	8.84	21.42	4.37

to verticillium wilt and to high ambient temperature, are grown. The performance of the principal varieties is summarized in Table 22.

Aleppo 40. Bred from a cross between Aleppo 1 and Acala SJ1, it was released in the late 1970s. It has a high level of tolerance to verticillium wilt, has a staple length of 1.333 inches and fibre strength of 21.4g/tex. It is grown mainly in the northeastern and central-northern areas and accounts for just over 50 percent of the total cotton area.

Aleppo 33/1. Selected from Acala SJ4, it is more tolerant to verticillium wilt than Aleppo 40 and is grown in the most wilt-infected areas of Hama Province. Staple length and fibre strength exceed those of Aleppo 40.

Rakka 5. Selected from the Russian variety, Tashkent 3, it is the earliest and most wilt-tolerant of all the current commercial varieties, although it has a lower ginning percentage than Aleppo 40. It is grown in areas of Rakka Province that are heavily infected with verticillium wilt.

Deir Ezzore 22. Selected from the American variety, Deltapine 41, it is adapted for Deir Ezzore Province where it is earlier, more heat-tolerant and higher-yielding with better fibre characteristics than other varieties.

PESTS

Pest infestation levels on cotton and losses of yield and quality are relatively low in Syria and consequently the need for pest control interventions is also low. No new pests of cotton have been introduced in the country in recent years.

Insects

Earias insulana *(Boisd.) (Lepidoptera: Noctuidae).* Spiny bollworm is the most important pest of cotton in Syria, especially in Deir Ezzore and Hama provinces of the eastern and central-northern regions respectively.

Helicoverpa armigera *(Hb.) (Lepidoptera: Noctuidae).* American bollworm is less important than spiny bollworm; it is distributed throughout the country, but reaches economic threshold levels on cotton mainly in the eastern and central-northern regions.

Agrotis ipsilon *(Hfn.) (Lepidoptera: Noctuidae).* Greasy, or black, cutworm may cause some losses to seedlings but this is not normally of economic significance because of high seeding rates. It is most prevalent in the central-northern and eastern regions.

Creontiades pallidus *(Ramb.) (Hemiptera: Miridae).* Cotton shedder bug is second only to spiny bollworm in economic importance on cotton in Syria, especially in Deir Ezzore Province in the eastern region.

Spodoptera exigua *(Hb.).* The lesser armyworm, or greenworm, is not usually an important pest, although there are occasional outbreaks, as in 1985 in the eastern region.

Bemisia tabaci *(Gen.) (Hemiptera: Aleyrodidae).* Cotton whitefly is not a serious pest at present, although it has the potential to become one. It is distributed throughout the cotton areas and is usually heavily parasitized by Aphelinids.

Sucking pests
Aphid, jassid, thrips and red spider are widely distributed but only occasionally reach pest status. Aphid and thrips are mainly seedling pests.

Diseases
Verticillium wilt. Verticillium wilt, caused by *Verticillium dahliae* (Kleb.) (formerly attributed to *V. albo-atrum* in Syria) is now less serious following the introduction of tolerant varieties such as Aleppo 40, Aleppo 33/1 and Rakka 5, and the rotation of cotton with wheat.

Seedling damping-off. There are several causal agents of damping-off at the seedling stage, including *Rhizoctonia solani* Kuhn, *Fusarium* sp. and *Alternaria* sp. Control is by seed dressings, either a combination of quintozene (PCNB) with a mercury compound or thiram, or carboxin alone.

Weeds
Farmers in Syria generally keep cotton fields very free of weeds by hoeing after germination. About 90 percent of the cotton area is treated with pre-emergence herbicides, usually trifluralin (incorporated into the soil at a rate of 2.5 litres of formulation per hectare). During the growing season fluazifop-P-butyl may be used as a spot treatment at 2.5 litres of formulation per hectare to control perennial grass weeds such as Bermuda grass (*Cynodon dactylon* (L.) Pers.) and Johnson grass (*Sorghum halepense* (L.) Pers.).

The main weed species in cotton are: the grass weeds (Gramineae) *Cynodon dactylon* (L.) Pers., *Echinochloa colona* (L.) Link, *Setaria italica* (L.) P. Beauv. and *Sorghum halepense* (L.) Pers.; and the broad-leaved weeds *Amaranthus retroflexus* L. (Amaranthaceae), *Xanthium brasilicum* Vell. (Compositae), *Convolvulus arvensis* L. (Convolvulaceae), *Chenopodium album* L. (Chenopodiaceae), *Portulaca oleracea* L. (Portulacaceae) and *Solanum nigrum* L. (Solanaceae).

CONTROL MEASURES
Chemical control
The use of insecticides is based on economic threshold levels (ETLs). Cotton fields are scouted each week by the extension services and the

resulting insect counts reported to the plant protection services in the provincial departments of agriculture. A committee of three officers then takes decisions on the need to spray. In a normal season only one pesticide application is needed, two applications are exceptional. When two or more applications are needed, different insecticide groups are used for each spray to reduce the risk of insecticide resistance. The current ETLs for the principal pests are:

- bollworm, live larvae on five out of 100 plants;
- cutworm, three larvae per square metre;
- armyworm, ten to 15 larvae per 100 plants;
- aphid, ten infested plants per 100;
- red spider, 10 to 20 percent infested leaves (at least three moving mites per leaf);
- shedder bug, ten bugs per 50 sweeps (until 70 percent boll formation).

Table 23 shows the area in hectares sprayed against the various pests in the eight seasons between 1986 and 1993. In these years the percentage of the total cotton area sprayed ranged from 2 to 19 percent, averaging about 7 percent per year. Small areas are sprayed with backpack sprayers or with tractor-drawn or tractor-mounted sprayers. When pest outbreaks are more widespread aerial application is carried out by state agencies. There is no local or foreign private-sector involvement in this.

The danger of whitefly becoming a key pest through the excessive use of insecticides early in the season to control spiny and American bollworm has led to the raising of the ETLs for bollworms from 2 to 5 percent and to the prohibition of insecticide use against whitefly itself. Table 24 lists the main insecticides used on cotton in Syria, with their target pests.

Legislative measures

There are a number of cultural and other control measures that farmers are obliged by the state to observe on their cotton. These measures include:

- Farmers must spray when advised to do so by the extension services; failure to comply can result in the spraying being done by the extension services, who then charge the farmer double the cost of the application.
- The use of insecticides against whitefly is banned; whitefly control is left to parasitoids and predators.

TABLE 23
Areas sprayed against target pests (in hectares)

Insects	1986	1987	1988	1989	1990	1991	1992	1993
Greenworm	308	1 573	2 127	159	309	-	-	53
Bollworms	113 82	11 224	368	751	873	2 086	8 559	5 218
Sucking insects	5 696	4 970	1 148	9 713	780	5 422	732	358
Red spider mite	1 595	2 743	129	660	402	2 636	186	422
Cutworm	0	3 776	2 872	650	70	25	62	-
Others	2 831	9	50	327	41	390	-	-
Total area sprayed *(ha)*	**21 721**	**24 292**	**6 690**	**12 290**	**2 475**	**7 715**	**9 539**	**6 051**
Cotton area	144 227	128 687	171 000	158 000	156 350	170 000	212 000	196 000
Percentage of cotton area sprayed	15	19	4	8	3	5	4	3

TABLE 24
Insecticides used in theSyrian Arab Republic

Insecticide and formulation	Target pests
Binapacryl wp	Mites
Carbaryl wp	American bollworm
Chlorobenzilate ec	Mites
Chlorpyrifos dp	Bollworms, armyworm, aphid
Cypermethrin ec	Bollworms, aphid, thrips
Cypermethrin ul	Bollworms, aphid, thrips
Deltamethrin ec	Bollworms, aphid, shedder bug
Deltamethrin ul	Bollworms, aphid, shedder bug
Endosulphan ec	Bollworms, armyworm, aphid
Endosulphan dp	Bollworms, armyworm, aphid
Formothion ec	Shedder bug, aphid, thrips
Methomyl wp	Sucking pests
Monocrotophos ec	Sucking pests
Pirimiphos dp	Bollworms
Triazophos ul	Bollworms, armyworm

- Only recommended varieties may be grown in any particular area.
- The last date for sowing cotton is set, usually for mid-May. The aim of this is to give the cotton time to set its main crop before bollworm populations build up in September, to permit boll maturation to be completed before the temperatures fall and to harvest as much of the crop as possible before autumn rains begin in October.
- Crop residues must be destroyed by a set date. Herding sheep on harvested fields, to graze remaining bolls and leaves, is a common practice which destroys bollworms and whitefly.
- An optimum plant population of 8 000 to 12 000 plants per hectare must be achieved.
- The use of nitrogenous fertilizers is regulated to prevent excessive vegetative growth.
- The recommended irrigation practices must be followed in order to avoid both water stress in the critical period (July to August) and water waste.
- Weeds must be controlled on uncultivated land and whitefly host plants must be destroyed in the non-cotton season.

Biological control

Parasitoids and predators are important in regulating cotton pest populations in Syria and the use of ETLs ensures that insecticides are not used unnecessarily, thus preserving the populations of beneficials, especially early in the season.

Varietal resistance

The Cotton Bureau screens new varieties for resistance to verticillium wilt. This has been a principal objective of the cotton breeding programme since the 1960s and now nearly all the main commercial varieties are wilt-resistant. So far there has been no breeding for insect resistance.

INFRASTRUCTURAL SUPPORT FOR COTTON IPM
Government support
Research. The Cotton Bureau conducts the bulk of cotton research in Syria, either on its own or in collaboration with other research organizations. Weed

research is carried out in collaboration with the Directorate of Scientific Agricultural Research and insect research with the Directorate of Plant Protection and the University of Aleppo. There is also collaboration with the Directorate of Soils, the Directorate of Irrigation and Water Use, the General Mechanization Organization and the universities. Ideally the Cotton Bureau should be able to cover all the areas itself, including cotton breeding, agronomy, physiology, plant pathology and entomology. The Cotton Bureau is particularly weak in entomology and IPM and needs overseas training for a team of specialists in these areas.

Plant protection services. The Directorate of Plant Protection maintains units in each provincial department of agriculture where they collaborate with cotton and extension specialists in providing farmers with advice, monitoring pest levels and controlling pests if necessary. The various directorates in MAAR collaborate with the Plant Protection Directorate to produce annual estimates for requirements of all types of pesticides. The necessary pesticides are then imported by a Ministry of Economics organization, GEZA, following authorization by the pesticides committee. The Agricultural Cooperative Bank distributes pesticides to its branches, to peasant unions and the agricultural engineering syndicate's pesticide stores. However farmers can only obtain pesticides on prescriptions issued by the extension units. Changes to ETLs, dosage rates and use regulations are decided at the annual Cotton Congress and passed to the Plant Protection Directorate for implementation and dissemination.

Extension services. As well as providing farmers with information and advice on cotton growing, the extension services grow demonstration plots. In 1991 for example, over 400 extension demonstrations were held.

Private agencies
The private sector is permitted by MAAR to import part of the annual pesticide requirements. However farmers still need prescriptions from the extension units to be permitted to purchase pesticides from private-sector dealers.

ASSISTANCE TO COTTON IPM
Government programme

The National Programme for Cotton in Syria (1984 continuing) has received support from the UNDP/FAO Programme SYR 84/001 (Improvement of Cotton Production in Syria 1983-1988). Under this programme there was provision for 50 working days of consulting time for an IPM specialist at the Cotton Bureau. The consultancy was filled by Mr O.G. Beingolea who reviewed previous research and commented on various aspects of the results. His findings were included in an UNDP/FAO technical report (Beingolea, 1986).

Beingolea's report highlighted differences in the results obtained by Stam (see the annex on p. 146) and the Cotton Bureau on the status of the cotton shedder bug, *Creontiades pallidus,* as a pest of cotton in Syria, and consequently the level at which the economic threshold levels should be set. Stam's research indicated that the ETL should be seven bugs per 50 sweeps. Populations about that level reduced yield through loss of or damage to fruiting points and this in turn resulted in excessive growth of the cotton plant.

Experiments carried out by the Cotton Bureau at Deir Ezzore between 1983 and 1985 showed that on plots sown at different times all plants showed heavy bud shedding that coincided with the period of high temperatures, irrespective of sowing date. These experiments also showed that excessive growth was correlated with heavy shedding. The same excessive growth could be induced in high plant populations, by heavy applications of nitrogen and by excessive irrigation. Experiments carried out in 1986 compared two plant populations at three levels of protection against shedder bug. The highest plant population receiving the highest level of protection (insecticide sprays at 14-day intervals), still exhibited rank growth and excessive shedding. In fact heavy shedding on all the plots, irrespective of the treatment, indicated the importance of high temperatures in inducing shedding.

It is now the view of the Cotton Bureau that the main cause of shedding is likely to be high temperatures, the widespread occurrence of this phenomenon compared with the patchy distribution of shedder bug lending support to this conclusion. Other experiments by the Cotton Bureau showed no difference in yield between plots receiving weekly insecticide sprays and unsprayed plots with peak shedder bug populations of up to 170 bugs per 50 sweeps.

These results reopen the question of how serious a pest shedder bug really is and at what level the ETL should be set. Further support for downgrading shedder bug's pest status and consequently raising the ETL is the observation that even though anthers may be damaged at the bud stage, flowers still open and may go on to form bolls, although this requires confirmation. Also, although young, green bolls may be punctured by shedder bug feeding, this does not necessarily preclude them from eventually maturing into ripe bolls. In other words, the amount of economic damage even a heavy infestation of shedder bug actually causes is in doubt. Heat-induced shedding would appear likely to subsume shedding caused by the bug.

Donor agency programme
The report of Dr Pieter Stam, the FAO entomologist/biological control expert assigned to the FAO/UNDP Project 0108/76/003 IPM/Cotton Pest Management in the Syrian Arab Republic 1979-83 is summarized in the annex on p. 146.

ACHIEVEMENTS AND IMPLEMENTATION
As a result of various research programmes cotton IPM in Syria has now reached a stage where:
* the main pests have been identified;
* economic threshold levels have been set for these pests;
* parasitoids and predators have been surveyed and identified;
* insecticides have been screened and those appropriate for effective pest control in IPM systems selected;
* two new varieties have been released; Rakka 5 with high levels of resistance to verticillium wilt and Deir Ezzore 22 with high heat tolerance. They have replaced Aleppo 40 in Rakka and Deir Ezzore provinces (eastern region).

KEY PESTS

The main pests of cotton in Syria are: spiny bollworm *(Earias insulana)*, American bollworm *(Helicoverpa armigera)*, shedder bug *(Creontiades pallidus)*, lesser armyworm*(Spodoptera exigua)*, whitefly*(Bemisia tabaci)* and sucking pests (aphid, thrips, jassid and red spider).

RECOMMENDATIONS FOR ALLEVIATING PROBLEMS

The IPM approach needs to be effectively implemented and refined through further research. Research topics which need pursuing include:

- *Spiny bollworm*: development of life tables to determine mortality at various stages; and investigation of non-chemical methods of control including the use of pheromones for mating disruption and mass trapping, the use of microbial pesticides *(Bacillus thuringiensis)* and the release of parasitoids *(Trichogramma)*.
- *American bollworm*: development of life tables; and investigation of non-chemical methods of control including viruses and pheromones and the release of *Trichogramma*.
- *Shedder bug*: investigations on the biology and life cycle of the bug; investigations on pest status and damage caused in relation to infestation level; effect of timing of attack on damage levels and yield; and alternative control methods, especially cultural methods.
- *Whitefly*: biology and host range and preference; insecticidal screening; and alternative control methods, especially cultural control, including trap cropping.
- *Biological control*: development of mass rearing techniques and facilities for augmentative releases of parasitoids and predators.
- *Varieties*: breeding early-maturing varieties and varieties resistant to pests.

KEY PERSONNEL INVOLVED IN COTTON PEST CONTROL IN SYRIA

Dr F. Khoury, Director of Research at the Cotton Bureau.

Dr K.I. Gomaa, entomology supervisor at the University of Aleppo.

Dr N. Salti, entomology supervisor at the University of Aleppo.

Mr A. Hammal, entomology researcher at the Cotton Bureau.

Mr A. Rashid, entomology researcher at the Cotton Bureau.

Mr K. Hazeim, entomology researcher at the Cotton Bureau.

Mr R. El-Lahm, Director of the Directorate of Plant Protection.

Dr K. El-Sheikh, entomology specialist at the Directorate of Plant Protection.

Mr Z. El-Yaffi, entomology specialist at the Directorate of Plant Protection.

Mr F. Moshref, Plant Protection Officer at the Department of Agriculture, Hassaka.

Mr A. Khamis, Plant Protection Officer at the Department of Agriculture, Deir Ezzore.

Mr A. El-Salem, Plant Protection Officer at the Department of Agriculture, Rakka.

Mr W. Aasi, Plant Protection Officer at the Department of Agriculture, Aleppo.

Mr A. Taiar, Plant Protection Officer at the Department of Agriculture, Hama.

Annex

SUMMARY OF RESULTS OF THE FAO/UNEP COTTON PEST MANAGEMENT PROJECT 1979-83

Pest population dynamics

Spiny bollworm, *Earias insulana,* is the main pest of cotton in Syria. There are six generations each year, of which four are on cotton, with the third and fourth doing the most damage.

- First generation – February to May;
- Second generation – May to June;
- Third generation – July to August;
- Fourth and fifth generations – August to October;
- Sixth generation – November.

American bollworm, *Helicoverpa armigera,* is the second-most important bollworm and, together with the cotton shedder bug, is the second-most important pest overall. There are usually about five generations each year.

- First generation – May;
- Second generation – late June;
- Third generation – mid-July;
- Fourth generation – early September;
- Fifth generation – late October.

Cotton shedder bud, *Creontiades pallidus,* is particularly important in Deir Ezzore province. Its feeding causes heavy shedding of pin squares and larger flower buds, especially between mid-June and early August. After August, damage is no longer of economic significance.

Other pests, particularly sucking pests, may occasionally cause economic damage. Whitefly, *Bemisia tabaci,* and red spider, *Tetranychus* sp., can attain major pest status when control by natural enemies breaks down.

The entomophagous fauna of Syria is substantial. The ladybird population is large in June and July. In Aleppo province for example, *Hyperaspis* sp. (Coleoptera: Coccinellidae) represented 62 percent of the predator number biomass, followed by another ladybird, *Coccinella undecimpunctata* L. with 5 percent. In the eastern region *Hyperaspis* was not found and *C. undecimpunctata* was the dominant predator with 11 percent of the predator number biomass.

A complex of hemipterous predators are also important as natural control agents in cotton, especially late in the season in August and September. During these two months in the years 1979 to 1982, *Orius laevigatus* (Fieber) (Hemiptera: Anthocoridae) represented 51 percent of the number biomass. *Campylomma diversicornis* Reuter (Hemiptera: Miridae) were also numerous, especially later in the season. Numbers of lygaeids and nabids were low. Spiders were numerous, representing 3.6 percent of the number biomass. *Chrysoperla carnea* (Stephens) (Neuroptera: Chrysopidae) was present in large numbers on cotton in June after which populations tended to decline. Larval numbers were generally lower than expected.

Effect of parasitoids and predators on cotton pests in Syria

Six experiments were conducted that demonstrated the part played by beneficial species in suppressing pest populations. When predator populations were reduced by about 27 percent, there was an increase of 6 percent in damaged fruiting points, resulting in a 19 percent reduction in yield (or 600 kg of seed cotton per hectare).

Observations on bollworm eggs showed that 10.3 percent of the eggs of spiny bollworm were parasitized in September and October, mainly by *Trichogramma semblidis* (Aurivillius) (Hymenoptera: Trichogrammatidae) although *T. chilonis* (Ishii) was occasionally observed. Hatching failure in the eggs of spiny and American bollworm was 32 percent and 25 percent respectively. Spiny bollworm larvae were parasitized by *Bracon brevicornis* (Wesmael) (Hymenoptera: Braconidae). Between early August and mid-September in Deir Ezzore province the mean rate was 2.3 percent, while larval mortality from unknown factors was 12.3 percent.

Populations of whitefly were heavily parasitized by *Eretmocerus mundus* Mercet (Hymenoptera: Aphelinidae) which on occasion caused the collapse of the whitefly population. Another aphelinid, *Encarsia lutea* (Masi), was occasionally observed in September. The parasitoids were extremely sensitive to insecticides; parasitism rates were only 7.4 percent in a treated field at Hawaj Deban in 1980 compared with 32.8 percent in an unsprayed field at Beni Taglib.

Yield losses to spiny bollworm

In an experiment conducted in 1982 a highly significant correlation was found between mean damage levels averaged over the season and yield of seed cotton per 10 m², as expressed by the equation:

$$Y = 4.690 - 0.091X$$

where X = mean damage and Y = expected yield of seed cotton in kg per 10 m².

The relationship between the mean percentage live spiny bollworm larvae and yields was also highly significant, expressed by the following equation:

$$Y = 4.019 - 0.41X$$

where X = percentage of live larvae and Y = expected yield of seed cotton per 10 m².

Use of these equations now enables accurate assessments to be made of the effect on yield of various levels of infestation.

Yields losses to cotton shedder bug

A series of experiments carried out between 1979 and 1982 showed that shedder bug can cause significant losses in yield. In 1979 and 1980 comparisons of shedding rates were made in different locations. The largest populations and the highest rates of shedding were observed in the Safira area in both years. In the observations a significant correlation was found between bug numbers and the number of flower buds with damaged anthers, expressed in the following equation:

$$Y = 17.069 + 3.541X$$

where X = number of bugs per 50 sweeps and Y = percentage of damaged buds.

In a cage experiment, in which cotton plants were exposed to large populations of shedder bug and compared to plants that were kept uninfested, the results showed a significant increase in plant height, node number and fruiting point shedding on the bug-infested plants. These plants showed a yield loss of 54.3 percent compared with the bug-free plants. Some 38.9 percent of this yield loss was caused by the shedding of buds and small bolls and 15.4 percent by damage to larger bolls. In the Safira area in 1982 it was found that a mean number of seven shedder bugs per 50 sweeps between early July and mid-August caused a yield loss of about 50 percent when infested fields were compared with bug-free fields.

The microbial insecticide *Bacillus thuringiensis* subsp. *kurstaki* (Dipel) was evaluated for spiny bollworm control in 1979. Although population buildup was delayed, the control of large populations was inadequate. The use of conventional chemical pesticides for spiny bollworm control did not always give satisfactory results. In 1980 the spiny bollworm population resurged after one application of mephosfolan at Hawaij Deban and the same resurgence occurred again in 1982 following one application of chlorpyriphos. Not only was spiny bollworm not controlled, but populations of beneficials were reduced, resulting in an outbreak of whitefly in 1980.

Integrated pest management (IPM) in Syria – where it is currently based on ETLs, the minimal use of insecticides and the encouragement of natural enemies – is giving good results, although whitefly represents a serious potential threat if great care in the use of insecticides is not sustained.

References

Barbandy, A.R. 1973. *Cotton insects in the Deir Ezzore province.* Report No. 40, Ministry of Agriculture and Agrarian Reform, Syrian Arab Republic. (In Arabic)

Beingolea, O.G. 1980. *Cotton protection through integrated pest control,* Consultancy report, TCP/SYR/001.

Beingolea, O.G. 1986. *Improvement of cotton production in Syria. Part III: integrated pest management.* Report for 1986, UNDP/FAO SYR/001. Rome, Italy, FAO.

Campion, D.G. 1980. *The use of sex pheromones for the control of cotton pests in Syria with particular reference to the spiny bollworm* Earias insulana *and the American bollworm* Heliothis armigera. Consultancy report, TCP/SYR/001.

Campion, D.G., McVeigh, L.J., Bettany, B.W. & Hunter-Jones, P. 1981. *The use of sex pheromones for the control of spiny bollworm and American bollworm in cotton growing in Syria.* Consultancy report, TCP/SYR/001.

Cotton Bureau. 1988. *Cotton Congress recommendations for 1989 season.* Aleppo, Syrian Arab Republic, Cotton Bureau.

Cotton Bureau. 1989. *Cotton Congress recommendations for 1990 season.* Aleppo, Syrian Arab Republic, Cotton Bureau.

Cotton Bureau. 1990. *Cotton Congress recommendations for 1991 season.* Aleppo, Syrian Arab Republic, Cotton Bureau.

El-Khateeb, A. 1969. Studies on the biology of the spiny bollworm, *Earias insulana* Boisd. (Lepidoptera, Noctuidea). Entomology Department, Faculty of Agriculture, Cairo University, Cairo, Egypt. (M.Sc. thesis)

Elmosa, H. 1982. *Report of the FAO/UNDP Near East Inter-Country Programme for the development and application of integrated pest control in cotton growing.* FAO/UNDP/0108/76/03. Rome, Italy, FAO.

Farbrother, H.G. 1982. *The extent and the likely cause of shedding of fruiting forms in the cotton crop of Deir Ezzore province.* Consultancy report, TCP/SYR/001. 21 pp.

Green, J., Ismail, M.S. & Beingolea, O.G. 1986. *Improvement of cotton production in Syria.* Consultancy report, FAO/SYR/84/. Rome, Italy, FAO.

Hariri, G. 1981. The problems and prospects of *Heliothis* management in southwest Asia. *In:* W. Reed and V. Kumble, eds. *Proceedings of the International Workshop on Heliothis Management,* p. 15-20. 15 to 20 November 1981, ICRISAT Centre, Patancheru, India.

Khouri, F.I. 1975. *Crop rotation as a method for controlling verticillium wilt of cotton in Syria.* Scientific conference of cotton. Aleppo, Syrian Arab Republic, Cotton Bureau.

Khouri, F.I. 1990. *Cotton production in Syria.* Aleppo, Syrian Arab Republic, Cotton Bureau.

Khouri, F.I. & Alcorn, S.M. 1973. Influence of *Rhizoctonia solani* on the susceptibility of cotton plants to *Verticillium albo-atrum. Phytopathol.,* 63.

Khouri, F.I. & Alcorn, S.M. 1973. Effect of *Meloidogyne incognita acrita* on the susceptibility of cotton plants to *Verticillium albo-atrum. Phytopathol.,* 63.

Mabrouk, A.A.M.M. 1967. Biological and ecological studies on *Earias insulana* Boisd. Faculty of Agriculture, Cairo Univesity, Cairo, Egypt. (M.Sc. thesis)

Shami, A.R. 1972. Control of verticillium wilt of cotton in Syria: production of a tolerant cultivar, Aleppo 1. *Coton et Fibres Trop.*, 27: 389-391.

Shami, A., Shakkour, S. & Hammal, A. 1978. Release of a new variety, Aleppo 40 tolerant to wilt, with better qualities. *Pak. Cottons,* July 1978: 253-268.

Stam, P.A. 1979. *Bio-control as part of an integrated pest control programme on cotton in the Syrian Arab Republic, report for 1979.* FAO/UNDP/0108/76/03. Rome, Italy, FAO.

Stam, P.A. 1981. *The effect of predators and parasites on the population dynamics of* Earias insulana, *and the importance of* Creontiades pallidus *as a pest on cotton in the Syrian Arab Republic, report for 1981.* FAO/UNDP/0108/76/03. Rome, Italy, FAO.

Stam, P.A. 1981. *A review of entomological findings on cotton in the Syrian Arab Republic over the years 1979, 1980 and 1981.* FAO/UNDP/0108/76/03. Rome, Italy, FAO.

Stam, P.A. 1982. *Continued observations on the quantitative impact of predators and parasites on populations of* Earias insulana *and on economic thresholds for* Creontiades pallidus *and* Earias insulana *on cotton in the Syrian Arab Republic, report for 1982.* FAO/UNDP/0108/76/03. Rome, Italy, FAO.

Stam, P.A. 1983. *Cotton pest management in the Syrian Arab Republic.* Cotton Bureau, Aleppo. Near East Regional Programme Integrated Pest Control, FAO/UNDP/0108/76/03. Rome, Italy, FAO.

Stam, P.A. & Sabek, S. 1980. *Biocontrol and the importance of some insect species on cotton in Syria, report for 1980.* FAO/UNDP/0108/76/03. Rome, Italy, FAO.

Turkey

C. Mart, S. Karaat, F. Tezcam, A. Sagir, M. Ali Goven, A. Atac,
I. Kadioglu, V. Cetin and A. Kismir

INTRODUCTION

It is known that cotton growing and weaving existed in Anatolia as early as the eleventh century. Marco Polo wrote in the thirteenth century that Turks were growing cotton and weaving fabric out of the yarn. The main development of cotton in Anatolia began with the American Civil War (1861). At that time the British, concerned that they would no longer be able to obtain the raw material for their textile industry from the United States, launched a campaign to supply the Ottomans with free cottonseed, machinery and equipment at low prices. The Ottoman Government, in turn, proclaimed a series of orders in 1862 which subsidized cotton production. In 1863, some 300 tonnes of cottonseed were brought from Egypt to be distributed free of charge and in 1864 the first spinning mills were established in Adana, Tarsus, Mersin and Kirkagac. These measures secured a considerable increase in cotton production, which reached 51 345 bales in 1872.

In 1908 the proclamation of constitutional monarchy brought further subsidies to encourage cotton production and a decree to the effect that central Asian cottonseed, as well as that from the United States, would be exempt from custom duties. By 1914 cotton production had reached 160 000 bales.

Cotton production and the cotton industry have developed most since the establishment of the Republic. The importance given to research and the assistance provided to farmers helped cotton production to exceed 2 million bales in 1968 (1 bale = 215 kg).

Today Turkey grows approximately 700 000 ha of cotton and produces 610 000 tonnes of cotton lint, with a yield of 880 kg per hectare. Turkey therefore ranks seventh among the major cotton-growing countries (Table 25).

TABLE 25

Major cotton-growing countries in area, production, yield and consumption of cotton

Country	Area (*'000 ha*)	Production (*'000 tonnes*)	Lint yield (*kg per ha*)	Consumption (*'000 tonnes*)
India	7 448	2 075	279	1 880
People's Republic of China	5 907	4 530	766	4 270
United States	4 640	3 352	722	1 951
Pakistan	2 644	1 645	619	1 245
Brazil	1 985	664	335	756
Uzbekistan	1 688	1 406	832	203
Turkey	670	617	925	567
Turkmenistan	591	410	693	11
Egypt	396	310	791	320
World	**33 109**	**18 701**	**565**	**18 597**

Source: World Statistics Bulletin of the International Cotton Advisory Committee, 1993.

Cotton yields have increased significantly since the 1950s (Table 26) and in recent years have been well above the world average of about 550 kg of lint per hectare. This improvement in yield has been caused by a number of factors, including the use of high-yielding varieties with superior fibre characteristics and the attention now given to seed quality, land preparation, crop husbandry (including sowing, spacing, weed control, fertilizer use and irrigation practices), pest and disease control and the introduction of mechanized agriculture.

Cotton has an important place in Turkey's economy, as it heads the list of export crops and represents a considerable source of foreign currency. The main cotton-growing areas are in the Mediterranean, Aegean and southeast Anatolia regions; they can be grouped into six ecological zones (Table 27).

When the southeast Anatolia project is completed, it will be possible to irrigate a further 1 641 282 ha and it is expected that cotton will occupy between 14 and 43 percent of the newly irrigated land, potentially doubling the growing area in Turkey.

TABLE 26

Cotton area, production and yield in Turkey, 1932 to 1991

Year	Area *('000 ha)*	Production *('000 tonnes lint)*	Lint yield *(kg per ha)*
1932	158	20	120
1942	327	74	227
1952	675	165	244
1962	660	245	371
1972	760	544	715
1982	595	489	822
1983	605	522	863
1984	760	580	763
1985	660	518	785
1986	585	518	885
1987	586	537	916
1988	723	649	894
1989	669	621	928
1990	641	655	1 021
1991	599	559	935

TABLE 27

Cotton area, production and yield for the main cotton-growing areas in Turkey, 1991

Year	Area *('000 ha)*	Production *('000 tonnes lint)*	Lint yield *(kg per ha)*
Aegean	261 215	243 033	930
Marmara	2 000	888	444
Mediterranean	218 392	205 323	940
Northeast	4 664	3 451	740
Southeast	103 629	98 631	952
Mid-east	8 720	8 100	929
Total	**598 620**	**559 426**	**935**

The cultivation of cotton in Turkey has a long history, but it was not until the early 1950s, when outbreaks of Egyptian cotton leafworm appeared in the Cukorova region, that problems with pests became important. Organochlorine and organophosphorus insecticides were then used on a large scale for leafworm control, but this led to the emergence of other pest species and to problems with resistance. Integrated pest management (IPM) research projects were started in 1970 in the Aegean, in 1973 in the Mediterranean and in 1984 in the southeast Anatolia regions. These projects covered the biology, ecology and population dynamics of both pest species and natural enemies, as well as methods of control. The results of the projects were reviewed and IPM programmes, including cropping techniques and cultural control methods, were presented to the extension services for each region.

CULTURAL PRACTICES
Land preparation
Methods of land preparation for cotton vary depending on the previous crop and on whether cotton is to be grown on the flat or ridges. All fields are ploughed to a depth of 23 to 30 cm, however.

Irrigation
Variations in local climate, soil characteristics, methods of irrigation and times of sowing cause differences in the water consumption levels of Turkey's cotton crop. In the Aegean and Mediterranean regions four to six irrigations are given to the crop, while in the southeast Anatolia region seven to 11 irrigations are applied. In the latter region high temperatures coupled with low relative humidity mean higher water requirements than elsewhere. Recommendations based on research are for the first irrigation to be applied 45 to 55 days after sowing, with subsequent irrigations at 15-day intervals. The timing of the first irrigation is indicated by the appearance of the plant; leaves are darker green at midday and the main stem shows a reddish coloration to within 12-15 cm of the growing point. The quantity of water given at an irrigation can vary between 90 and 125 mm according to growth stage. Furrow irrigation is the most common system, although experiments have been carried out on various methods of overhead irrigation.

Fertilizer

Recommendations, based on research, for the Aegean and Mediterranean regions are for 120 to 160 kg of nitrogen (N) per hectare and 60 to 70 kg of phosphorus (P) per hectare. In the southeast Anatolia region the recommendation is for 130 kg N per hectare and 70 kg P. The application of N is normally split, half at the time of sowing and half just before the first post-sowing irrigation. N is placed at a depth of 8 to 10 cm, 10 to 15 cm from the cotton plants. Phosphate fertilizer is applied before sowing.

Rotation

Normally cotton is grown in rotation with wheat, but since the 1980s other crops have been introduced into the rotation, with soybean or maize being sown as a second crop after wheat and before cotton. These crops, or cucurbits, may be sown as the first crop before cotton. Cotton can be grown on the same land for three to four years if it is economically advantageous. In the Mediterranean region the rotation is wheat-maize or soybean-cotton-wheat. Cotton is hand-picked in Turkey.

VARIETIES

The different cotton-growing areas grow locally adapted varieties as follows:
- *Aegean region:* Nazilli 84 (the main variety), Nazilli 87, Nazilli 66-100, Ege 69, Del Cero, Ege 7913 and Gossypol 86.
- *Mediterranean region:* Cukurova 1518 (main variety), Carolina Queen, Deltapine 15/21, Deltapine 20, Deltapine 50 and Deltapine 61.
- *SoutheastAnatolia region:* Sayar 314, Maras 92 (Sat-92), Ersan 92 (Kat-92), McNair 235, McNair 612 and Stoneville 825.

Varieties grown in the Mediterranean region are early-maturing and resistant to whitefly. Those grown in the other two regions are resistant to verticillium wilt.

PESTS
Insects and mites

Insect and mite pests are most serious in the Mediterranean region (Table 28). In the southeast Anatolia region, where future expansion of cotton

TABLE 28

Relative importance of the main insect and mite pests in the cotton-growing regions of Turkey

Species	Cotton-growing regions		
	Mediterranean	Aegean	Southeast Anatolia
Bemisia tabaci (Gen.)	***	*	*
Aphis gossypii (Glov.)	***	**	**
Asymmetrasca decedens Paoli	**	**	*
Empoasca decipiens Paoli	*	*	*
Thrips tabaci Lind.	*	**	***
Frankliniella intonsa Tryb.	**	*	*
Nezara viridula (L.)	*	*	*
Helicoverpa armigera (Hb.)	***	***	***
Spodoptera littoralis (Boisd.)	***	*	*
S. exigua (Hb.)	**	**	**
Pectinophora gossypiella (Saund.)	*	*	
Agrotis segetum (Schiff.)	**	**	**
A. ipsilon (Hfn.)	**	**	**
Lygus gemellatus (H.-S.)	*		
Creontiades pallidus (Ramb.)	*		
Tetranychus urticae (Koch.)		***	***
T. cinnabarinus (Boisd.)	***		

*** = Main pest. ** = Secondary pest. * = Unimportant pest.

production is planned, pests are less serious as natural control is still effective.

Bemisia tabaci *(Gen.) (Hemiptera: Aleyrodidae).* Since the outbreaks of 1974, whitefly has become the major pest of cotton, especially in the Cukorova area of Adana in the Mediterranean region, although parts of Antalya and Hatay also have problems comparable with those of Adana. Whitefly is locally distributed in the Aegean region and occurs only at low densities in the southeast Anatolia region. Whitefly populations reach their maximum levels between mid-July and the end of August, when boll formation is at its peak, and most damage is done at this time. Crop losses of up to 76 percent have been recorded, but vary with season and the intensity of the attack.

Helicoverpa armigera *(Hb.) (Lepidoptera: Noctuidae)*. American bollworm is a major pest in all the cotton-growing areas. In the Mediterranean region there are five generations each year, three of which are on cotton.

Aphis gossypii *(Glov.) (Hemiptera: Aphididae)*. The status of aphid has changed in recent years. Until 1984 it was considered an early-season pest, but more recently in the Mediterranean region very high population levels have developed following the first post-sowing irrigation. In the Aegean region aphid was considered the main pest until 1989, but now it is found only occasionally and usually late in the season. In the southeast Anatolia region it is also found only occasionally. It has developed resistance to a number of insecticides.

Tetranychus cinnabarinus *(Boisd.) and* **T. urticae** *Koch. (Acari: Tetranychidae)*. The cotton red spider *(T. cinnabarinus)* is found in the Adana, Mersin and Hatay districts of the Mediterranean region. It has become highly resistant to a number of acaricides. The two-spotted red spider *(T. urticae)* is found in the Aegean and southeast Anatolia regions where it has not developed acaricide resistance. Both species are considered major pests in their areas of distribution.

Spodoptera littoralis *(Boisd.) (Lepidoptera: Noctuidae)*. Egyptian cotton leafworm is a major problem only in the Mediterranean region, especially in Adana, Mersin, Antalya and Hatay provinces.

Thrips tabaci *Lind. (Thysanoptera: Thripidae)*. Thrips is a serious early-season pest, particularly in the southeast Anatolia and Aegean regions. It is of less importance in other cotton-growing areas.

Frankliniella intonsa *Tryb. (Thysanoptera: Thripidae)*. This has been a pest in the Mediterranean region only since 1984.

Asymmetrasca *(syn.* **Empoasca***)* **decedens** *Paoli and* **E. decipiens** *Paoli (Hemiptera: Cicadellidae).* Jassid has become a particularly damaging early-season pest in the Mediterranean region in recent years.

Other pests. Other pests include *Nezara viridula* (L.) (Heteroptera: Pentatomidae), shield bug; *Spodoptera exigua* (Hb.) (Lepidoptera: Noctuidae), lesser armyworm; *Pectinophora gossypiella* (Saund.) (Lepidoptera: Gelichiidae), pink bollworm; *Agrotis segetum* (Schiff.) and *A. ipsilon* (Hfn.) (Lepidoptera: Noctuidae) cutworms; *Lygus gemellatus* (H.-S) (Hemiptera: Miridae), plant bug; and *Creontiades pallidus* (Ramb.) (Hemiptera: Miridae), shedder bug. The comparative status of these, and the major pests, in the different cotton-growing areas, is shown in Table 28.

The Mediterranean region suffers most from pest attack. Table 29 shows the total hectares in 1986, 1988 and 1990 where the major pests reached economc threshold levels.

Diseases

The most important diseases of cotton are verticillium wilt, *Verticillium dahliae* Kleb., and a complex of pathogens causing seedling damping-off,

TABLE 29

Area of cotton infested by major pests in the Mediterranean region (Adana, Mersin and Antalya), 1986, 1988 and 1990

Pest	Area *(ha)*		
	1986	1988	1990
Whitefly	121 300	171 900	124 600
American bollworm	127 700	123 700	119 100
Red spider	45 800	37 400	56 300
Aphid	168 800	193 000	140 600
Leafworm	75 600	37 800	9 300
Thrips	27 000	21 300	29 200
Jassid	36 200	22 300	30 800
Total cotton area	**207 030**	**259 450**	**181 604**

Source: Annual reports of plant protection extension services in Mediterranean region.

including *Rhizoctonia solani* Kuhn., *Alternaria* spp., *Fusarium* spp., *Macrophomina* spp. and *Xanthomonas campestris* pv. *malvacearum* (E.F. Smith) Dye. Occasionally there is a failure of some bolls to split. These disease problems affect all cotton-growing areas.

Weeds

The most important weed competitors of cotton in Turkey are listed in Table 30. Losses to weeds in Turkey are somewhat higher than the world average of 5 to 6 percent. The most important weeds, on the basis of their

TABLE 30

Major weeds of cotton in Turkey

Weed type	Weed species
Annual grass weeds (Gramineae)	*Digitaria sanguinalis* (L.) Scop. *Echinochloa colona* (L.) Link. *E. crus-galli* (L.) P. Beauv. *Eragrostis cilianensis* (All.) Lut. *Setaria verticillata* (L.) P. Beauv.
Annual broad-leaved weeds	*Amaranthus albus* L. (Amaranthaceae) *A. deflexus* L. (Amaranthaceae) *A. retroflexus* L. (Amaranthaceae) *A. gracilis* Desf. (Amaranthaceae) *Chenopodium album* L. (Chenopodiaceae) *Chrozophora tinctoria* (L.) Raf. (Euphorbiaceae) *Datura stramonium* L. (Solanaceae) *Euphorbia macroclada* Boiss. (Euphorbiaceae) *E. prostrata* Ait. (Euphorbiaceae) *Heliotropium europaeum* L. (Boraginaceae) *Hibiscus trionum* L. (Malvaceae) *Physalis lanceifolia* L. (Solanaceae) *Portulaca oleracea* L. (Portulacaceae) *Solanum nigrum* L. (Solanaceae) *Tribulus terrestris* L. (Zygophyllaceae) *Xanthium spinosum* L. (Compositae) *X. strumarium* L. (Compositae)
Perennial grass weeds (Gramineae)	*Cynodon dactylon* (L.) Pers. *Imperata cylindrica* (L.) Raeuschel *Paspalum paspalodes* (Michx.) Scribner *Phragmites australis* (Cav.) Steud. *Sorghum halepense* (L.) Pers.
Perennial broad-leaved weeds	*Alhagi pseudalhagi* (Bieb.) Desv. (Leguminosae) *Convolvulus arvensis* L. (Convolvulaceae) *Glycyrrhiza glabra* L. (Leguminosae) *Hypericum perforatum* L. (Hypericaceae) *Prosopis farcta* (Banks & Sol.) Macbride (Leguminosae)
Perennial sedge (Cyperaceae)	*Cyperus rotundus* L.

distribution, density and the control difficulties they present, include *Sorghum halepense* (L.) Pers, *Cyperus rotundus* L., *Solanum nigrum* L., *Xanthium strumarium* L., *Setaria verticillata* (L.) P.B., *Echinochloa colona* (L.) Link., *Portulaca oleracea* L., *Convolvulus arvensis* L. and *Amaranthus albus* L.

Stickiness
The contamination of cotton lint by extraneous and sticky substances is a major problem, although it varies from region to region. It was first observed following two outbreaks of whitefly in 1974 in the Cukorova area (Table 31).

TABLE 31
Extraneous substances and stickiness levels of Turkish cotton

Region	Extranuous substance *(%)*				
	Insignificant	Medium	Significant	Sticky	No. of samples
Cukorova	43	19	38	31	16
Aegean	82	9	9	9	22
Antalya	45	41	14	14	11
Southeast Anatolia	71	29	0	14	7

Source: Survey on Cotton Contamination, International Textile Manufacturers' Federation.

CONTROL METHODS
Chemical control
Chemicals are the most important means of pest, disease and weed control in cotton in Turkey. Pest control accounts for 25 percent of production costs. Control methods and economic threshold levels for the main pests are summarized in the following; currently recommended insecticides and acaricides are listed in Table 32, herbicides in Table 33 and seed dressings in Table 34.

Insects and mites. Whitefly. Soil-applied granular formulations are widely used for whitefly control. The timing of application depends on whitefly

TABLE 32

Insecticides and acaricides used on cotton in Turkey, with target pests

Insecticide/acaricide common name	Whitefly	Aphid	Jassid	American bollworm	Leafworm	Lesser armyworm	Cutworm	Spider mite and Lygus	Shedder bug	Spiny bollworm
Aldicarb	*									
Acrinathrin								*		
Amitraz	*									
Azinphos-methyl		*								*
Bifenthrin	*			*				*		
Benfuracarb		*								
Bromophos		*								
Bromopropylate								*		
Buprofezin	*									
Carbaryl			*	*						
Carbosulphan		*								
Chlorfluazuron					*					
Chlorpyrifos		*			*		*			
Clofentezine								*		
Cyfluthrin				*		*				
Lambda-cyhalothrin				*						
Cypermethrin	*			*	*					
Deltamethrin	*			*	*					
Diafenthiuron	*	*						*		
Dialifos	*							*		
Diazinon		*	*							
Dicofol								*		
Dicrotophos		*	*			*				
Dimethoate		*	*					*	*	
Dioxabenzofos				*						
Endosulphan		*		*		*	*			
Fenbutatin oxide								*		
Fenitrothion		*								*
Fenpropathrin	*					*		*		
Fenthion		*		*						
Fenvalerate	*			*	*					
Flucythrinate				*	*					
Formothion	*	*						*	*	
Furathiocarb		*								
Hexaflumuron					*					
Hexythiazox								*		
Malathion		*	*			*				
Mephosfolan		*				*		*		

(continued)

TABLE 32 (continued)

Insecticide/acaricide common name	Whitefly	Aphid	Jassid	American bollworm	Leafworm	Lesser armyworm	Cutworm	Spider mite and Lygus	Shedder bug	Spiny bollworm
Methamidophos		*			*			*	*	
Methomyl				*						
Monocrotophos	*	*						*	*	
Omethoate		*								
Oxydemeton-methyl		*						*	*	
Permethrin	*			*						
Phosalone		*			*			*		
Phospholan		*			*					
Pirimiphos-methyl	*									
Profenofos		*		*						
Propargite								*		
Pyridaphenthion	*									
Quinalphos		*								
Tetradifon								*		
Thiodicarb		*		*	*					
Thiometon		*								
Triazophos	*							*		
Buprofezin + lambda-cyhalothrin	*									
Bifenthrin + amitraz	*									
Fenpropathrin + pyriproxyfen	*									
Hexythiazox + fenpropathrin								*		
Methomyl + diflubenzuron					*					
Profenofos + cypermethrin				*						

numbers, plant growth stage and the persistence of the selected insecticides. Normally the application is made before the second irrigation, if irrigation has started early, or before the first irrigation if irrigation has started late. Foliar sprays are based on an economic threshold level (ETL) of five adults or ten larvae and/or pupae, per leaf. At least two sprays are applied, at seven- to 14-day intervals. Two applications are normally sufficient in Hatay, Gazientep, Kahramanmarab and Antalya, but in Adana and Mersin three sprays may be required. Aerial spraying is considerably less effective against

TABLE 33
Dosage and application time of principal herbicides used in Turkey

Herbicide and formulation	Dosage *(kg ai per ha)*
Presowing	
Alachlor 48% ec	1.92
Dinitramine 25% ec	0.50
Ethalfluralin 50% ec	0.99
Fluridone 50% sc	0.20
Metolachlor 50% ec	2.00
Paraquat 20% ec	0.80
Trifluralin 48% ec	0.96
Pre-emergence	
Fluometuron 80% wp	2.40-3.20
Linuron 50% wp	0.75-1.00
Pendimethalin 33% ec	1.32
Prometryn 50% sc	2.00
Prometryn 80% wp	2.00
Postemergence	
Dalapon 85% wp	7.22
Fenoxaprop 12% ec	0.24
Fluazifop-butyl 25% ec	0.25
Fluazifop-p-butyl 12.5% ec	0.125
Haloxyfop-etotyl ec	0.125
Quizalofop 50% ec	0.37-0.50
Sethoxydim 20% ec	0.60-0.80

larvae and pupae than ground spraying, especially when ground sprayers are used that give good underleaf cover. Irrigation water and tall plants make ground application difficult but the advantages outweigh the disadvantages with this type of sprayer.

American bollworm. Chemical control is most effective when directed against early-stage larvae; older larvae become very difficult to control.

TABLE 34

Seed treatment chemicals used on cotton in Turkey

Seed treatment and formulation	Quantity *(per 100 kg cottonseed)*
Seedling root rot	
Carboxin + thiram 20.5 + 20.5% ec	500 ml
Carboxin + thiram 37.5 + 37.5% wp	500 g
Chloroneb 10% ds	2 000 g
Pencycuron + captan 20 + 50% ds	500 g
Quintozene 18% ds	2 500 g
Quintozene 75% ds	600 g
Tolclofos-methyl 10% ds	2 000g
Bacterial blight	
Bronopol 12% ds	600g
Mancozeb 60% wp	150 g
TCMTB 745g/ I ec	225 ml

Note: Quintozene + captan 10% + 10% dust as foliar treatment at 7 kg per ha is also used as a treatment against seedling root rot.

Light traps and pheromone traps are used by research and extension services to provide an early warning to farmers of likely bollworm infestation. Sprays are applied when pest monitoring indicates that the ETL of two larvae per 3 m of cotton row has been reached. Three sprays are applied in Adana and Mersin, two in Antalya and Hatay and only one in Gaziantep, Kahramanmarab and the southeast Anatolia region.

Aphid. The ETL is 50 percent of plants infested before the first square appears and, after squaring has commenced, an average of 25 aphids per leaf. The Adana and Mersin areas may require up to three applications, while Antalya, Gazientep, Hatay and Kahramanmarab may need only two sprays. Sprayers giving good underleaf cover are the most effective.

Spider mites. Spraying field margins is recommended for the two-spotted and red spider mites. Two sprays are needed in Adana and Mersin provinces, only one treatment being necessary in other areas. The ETL in July and August, when population levels may be high, is five mobile-stage red spider mites or 10 two-spotted spider mites per leaf. For optimum control sprayers giving good underleaf cover are recommended.

Leafworm. The ETL is two egg masses per 25 plants or five larvae per ten plants. Leafworm is a problem only in the Mediterranean region, in Adana, Mersin and Hatay provinces, where one or two sprays may be needed.

Other pests. Chemical control of both thrips is recommended when the infestation levels reach 15 percent. The ETL for jassid is ten nymphs or adults per leaf.

Weeds. Herbicides may be applied before sowing to control annual grass and broad-leaved weeds. Trifluralin is usually used for this application. Prometryn is used as a pre-emergence herbicide to control annual broad-leaved weeds. One of the most important weeds, *Sorghum halepense* (Johnson grass), is controlled with postemergence herbicides.

Seedling diseases. Cottonseed for sowing is dressed to give protection from *Rhizoctonia, Xanthomonas* and cutworms.

Legislative control.

The regulations concerning pest control and plant quarantine are contained in Law No. 6968 of 1957. These regulations cover the import, export and internal movement of all plant materials and their protection from pests and diseases, as well as the import, export, production, distribution and sale of all equipment and chemicals used in pest control. The same general law contains a by-law on pink bollworm on cotton (*Pectinophora gossypiella* Saund.) the object of which is to ensure that populations of pink bollworm larvae are minimized. The by-law sets out various rules to be followed in spinning mills and homes where cotton yarn is spun, as well as by cotton producers. Criteria for infested and uninfested cottonseed are defined, to be applied to oil mills, cottonseed producers and farmers. These regulations have been successful in ensuring that pink bollworm is not a problem in Turkey.

Cultural control

The whitefly problem can be reduced by early sowing, wide interrow spacing (this was increased from 80 to 120 cm) and the avoidance of excessive applications of irrigation water and nitrogen, which can increase whitefly populations without leading to an increase in yield. Insecticides are most effective against whitefly if applied after irrigating.

Biological control

Classical biological control is not practised on cotton in Turkey. Natural enemies do exercise control of many pests, however, and the pest management strategy is designed to enhance the role of beneficial species, especially through the selection of so-called "soft" pesticides and the timing of their application. In the southeast Anatolia region natural enemies are particularly numerous and effective in cotton fields. Some important parasitoids and predators of cotton pests are listed in Table 35.

Manual and mechanical control

Much weed control in cotton is by hand-hoeing. Saw-ginning and field sanitation are important in pink bollworm control.

Resistant varieties

Cultivars resistant to verticillium wilt include Nazilla 84 and Nazilla 87 and the hybrids Kat-64 and Sat-92. Cukurova 1518, being an early-season variety, is less susceptible to whitefly. In trials glabrous, okra-leaf and increased gossypol characters show promise for enhancing resistance to whitefly and American bollworm.

RESISTANCE TO PESTICIDES

Pest resistance to pesticides is particularly prevalent in the Mediterranean region, especially in Cukurova. Where resistance occurs farmers tend to increase dosages, thereby reducing profits, harming beneficial species and increasing the risk to the environment. Resistance to various pesticides occurs in the following major pests.

TABLE 35
Natural enemies of cotton pests occuring in Turkey

Order	Family	Species
Predators		
Coleoptera	Coccinellidae	*Coccinella septempunctata* L.
		C. undecimpunctata L.
		Adonia variegata (Goeze)
		Platynaspis luteorubra (Gz.)
		Propylaea quadridecimpunctata (L.)
		Exochomus nigromaculatus (Gz.)
		E. flavipes M.
		Hyperaspis quadrimaculata Red.
		Stethorus gilvifrons (Mulsant)
		Scymnus apetzi (Mulsant)
		S. quadriguttatus Fursch-Kreissel
		S. rubromaculatus
		S. punctillum
		S. levaillanti Mulsant
		S. bivulnerus Capra and Fursch
		S. araraticus Khnzorian
		S. flavicollis Redtb.
		S. pallipediformis Gunther
		S. interruptus (Gz.)
Heteroptera	Lygaeidae	*Geocoris pallidipennis* (C.)
		G. lineola (Rb.)
		G. megacephalus (R.)
		G. ater (L.)
		Piocoris erythrocephalus (P.-S.)
		P. luridus Fieb.
	Nabidae	*Nabis pseudoferus* Rm.
		N. rugosus L.
	Anthocoridae	*Orius niger* (W.)
		O. horwathi (Reut.)
		O. minutus (L.)
	Miridae	*Deraeocoris pallens* Reut.
		D. serenus Dgl. Sc.
		Campylomma diversicornis Reut.
Thysanoptera	Thiripidae	*Scolothrips longicornis* Priesner
	Aeolothripidae	*Aeolothrips fasciatus* L.
		A. collaris Priesner
		A. intermedius Bagnal
Neuroptera	Chrysopidae	*Chrysoperla carnea* (Steph.)
Diptera	Syrphidae	*Metasyrphus corralae* (Fabr.)
		Episyrphus balteatus (De Geer)
		Scava pyrastri (Linneaus)
	Cecidomyiidae	*Aphidoletes aphidimyza* Rondani

(continued)

TABLE 35 (continued)

Order	Family	Species
Parasitoids		
Hymenoptera	Aphelinidae	*Eretmocorus mundus* Mercet
		Encarsia sp.
		Prospaltella sp. nr. *aspidioticola* M.
	Elasmidae	*Elasmus platyedrae* Ferr.
	Ichneumonidae	*Hyposoter didymator* (Thbg.)
		Ichneumon sarcitorius L.
		Barylypa humeralis (Brauns)
		B. carinata (Brischke)
	Braconidae	*Habrobracon hebetor* Say.
		Lysophlebus fabarum Marshall
		L. confusus Tremblay and Eady
		Microbracon mellitor Say.
		Microplitis rufiventris Kok.
		Meteorus rubens
		Cotesia ruficrus Halliday
	Eulophidae	*Pnigalio soemius* (Walk.)
		Pediobius bruchicida (Rond.)
Diptera	Tachinidae	*Gonia bimaculata* Eied.
		Gonia sp.

Egyptian cotton leafworm. In 1973 monocrotophos and methamidophos ceased to be recommended for control of leafworm, followed by cypermethrin, deltamethrin, fenvalerate, chlorpyrifos and methomyl after 1986/87 when reductions in their effectiveness first became apparent in the Mediterranean region.

Spider mite. Systemic organophosphorus acaricides, first recommended in 1979, are now no longer recommended. Since 1986 the effectiveness of the specific acaricides, bromopropylate, dicofol, propargite and tetradifon has declined and is the subject of research in the Mediterranean region. Oxydemeton-methyl and dimethoate are no longer recommended in the Aegean region.

Whitefly. Following the outbreaks of whitefly in 1974 cypermethrin, fenvalerate, deltamethrin, pirimiphos-methyl and triazophos were used to excess, and resistance to these products developed. They were withdrawn from the list of recommended chemicals in the Mediterranean region in 1982. Amitraz and pyridaphenthion have become less effective since 1986.

American bollworm. In the Mediterranean region endosulphan and chlorpyrifos-ethyl ceased to be recommended for American bollworm control in 1975; deltamethrin, cypermethrin and fenvalerate in 1984; flucythrinate, permethrin and quinalphos in 1985; and in 1986 loss of effectiveness was observed in profenophos and methomyl.

Aphid. In the Mediterranean region a large number of insecticides have ceased to be recommended for aphid control because of resistance. In 1984 oxydemeton-methyl, phosphamidon, diazinon, formothion, pirimicarb, endosulphan, dimethoate, phosalone, azinphos-methyl, malathion, thiometon and fenitrothion were withdrawn; and in 1991 monocrotophos, profenophos, methamidophos, methomyl and mephosfolan were withdrawn. In the Aegean region oxydemeton-methyl and formothion have ceased to be recommended for aphid control.

IPM RESEARCH ON COTTON

Since 1990 the Ministry of Agriculture and Rural Affairs (MARA) has taken responsibility for all research on cotton pest, disease and weed control under a National Cotton IPM, Development and Training Project. Twelve subprojects on aspects of IPM are listed in the Annex on p. 184 which also shows the research institutes responsible and the principal investigators.

Before 1991 the three research institutes for the main cotton-growing areas ran independent IPM programmes, their main periods of activity being 1973 to 1984 for Adana PPRI; 1970 to 1981 for Bornova-Izmir PPRI; and 1984 to 1989 for Diyarbakir PPRI. Between 1973 and 1979 investigations were carried out on beneficial species in the Cukurova region. The winter behaviour and host range of whitefly were also studied. It was discovered that it could continue to reproduce in the winter and it was found on 47 different host plants. Between 1981 and 1984 further research on whitefly was carried out. Crop loss assessments showed that in the absence of chemical control whitefly could cause up to 67 percent loss in yield of cotton.

Re-evaluation of ETLs showed that they could safely be raised from 11 to 25 nymphs and adults for aphid and from six to ten whitefly larvae and pupae. Varietal resistance studies showed that the variety Carolina Queen, because of its short season, carried lower whitefly populations than other varieties. It was also found that pheromone traps could provide a reliable indication of when to commence spraying against American bollworm.

In the 1980s FAO supported IPM research in Cukurova through Project No. AG:DP/TUR/83/008. This project demonstrated that farmers could profitably implement the IPM approach to cotton control, incorporating suitable cultural, biological and chemical control methods.

In the Aegean region IPM research demonstrated that the key pests were aphid and red spider, with whitefly important in some localities. In the southeast Anatolia region it was found that aphid, two-spotted mite, tobacco thrips and bollworm could reach economic threshold levels but, because there had been less pesticide use than in other regions, natural enemies were still important in exercising control.

INFRASTRUCTURAL SUPPORT FOR COTTON IPM
Plant protection research institutes. There are seven institutes involved in pest control research in Turkey; four are crops research institutes and three are specialized plant protection research institutes (PPRIs). The latter are located at Adana, Bornova-Izmir and Diyarbakir. They conduct research on cotton pests, diseases and weeds, including pest biology and ecology. Until 1989 all candidate crop protection chemicals were tested by the PPRIs, but since then the private sector has been responsible for running its own trials. The results of these trials are assessed by a committee of the General Directorate for Plant Protection and Control of the Ministry of Agriculture and Rural Affairs (MARA). This committee then makes recommendations regarding the registration of pesticides based on its assessment of the trial results.

As well as conducting research, the PPRIs organize training courses and seminars for technical personnel in the plant protection extension services and publish technical bulletins and literature on crop protection aimed at farmers.

Research-extension links. Research results from the PPRIs and other research institutes are regularly reviewed, at institute level by institute research committees, later by plant pests research groups and ultimately by the Plant Protection Research Council, which takes the final decision on the use to be made of the research results. Scientific papers on crop protection research are published in MARA's Plant Protection Bulletin.

Where appropriate, the output of research is conveyed to the farmers through MARA's technical instructions, which include information on pests, diseases and weeds, their biology, incidence and host plants, the damage caused and control methods. The extension services use the material in these publications in the preparation of literature aimed at farmers, cooperatives and private companies. The importance of maintaining close links between research and extension is recognized and there are many cooperative projects between the two groups, including the operation of demonstration plots and the preparation of radio and television programmes.

Plant protection extension services. MARA functions at provincial and district levels through directorates and sub-directorates (Figure 4). In each province where cotton is grown the ministry forms mobile cotton teams which tour the villages in their areas of responsibility for four days each week. As well as giving farmers general advice and support these teams monitor pest levels and decide, on the basis of ETLs, the need for control interventions. In their work the mobile teams are assisted by district- and village-level staff. On the Friday of each week these plant protection staff meet and, together with experts from the research institutes, review the cotton and cotton pest situation. Farmers are kept informed of developments through the media, extension leaflets and personal contacts.

Pesticide application. Farmers are responsible for their own pest control operations in Turkey and must pay the costs of chemicals and equipment. There are no grants for chemicals and no equipment available for loan.

Farmers obtain pesticides from two main sources. The first source are cooperative unions, such as the Cukobirlik (Cukurova Cotton Producers' Union) which has branches in all the southeastern and Mediterranean

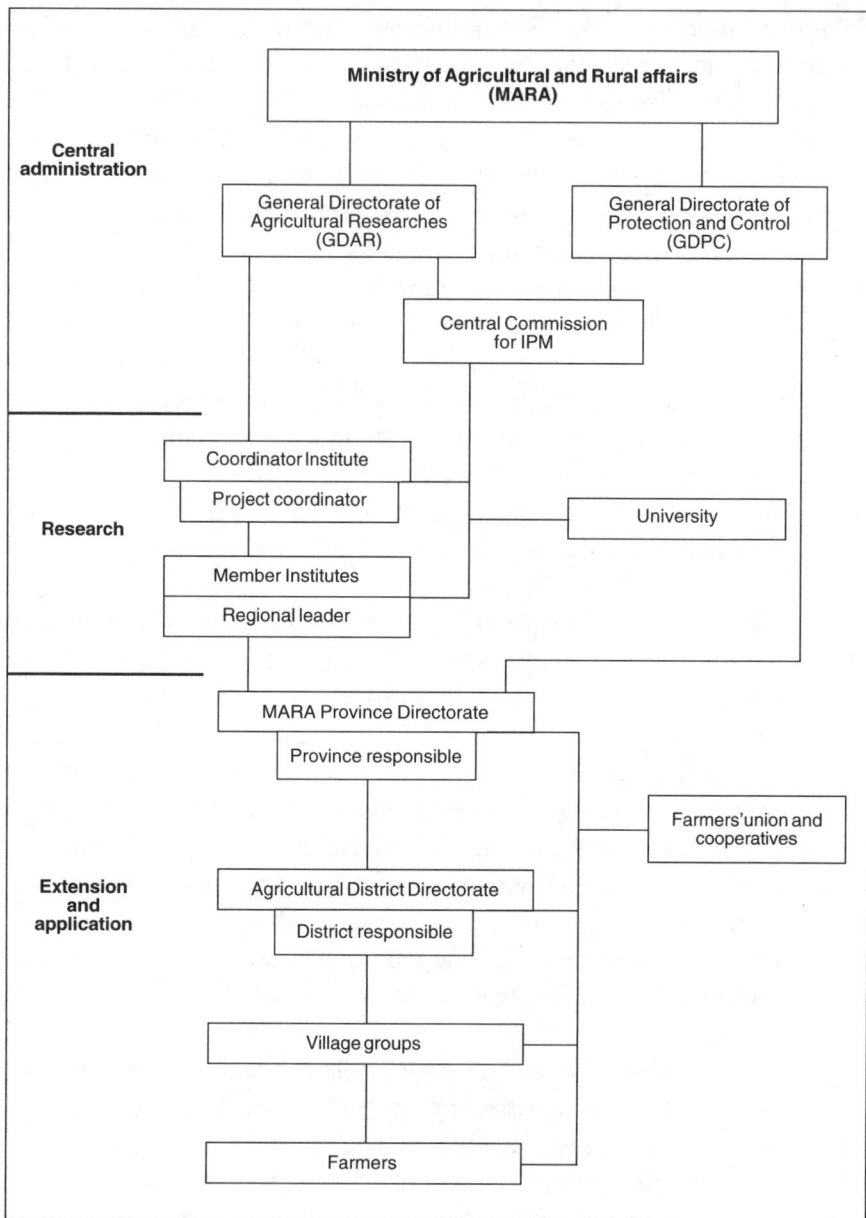

FIGURE 4
Organization of crop protection in Turkey under the Ministry of Agriculture and Rural Affairs

provinces (except Antalya), the Antbirlik (Antalya Cotton Producers' Union), the Tarip (Aegean Producers' Union) and the agriculture-sale credit cooperatives, which are located throughout the country. All these cooperatives have the status of state economic enterprises and purchase products by tender from agrochemical companies.

The second source of pesticides are the private-sector retailers. Retailers are regulated by the General Directorate of Plant Protection of MARA. They charge higher prices for pesticides than the cooperatives and farmers who are not cooperative members must rely on retailers for their needs. However the government refunds 20 percent of the cost of chemicals to these farmers.

Pesticides are applied to cotton using a wide range of equipment, including backpack sprayers, mist blowers, tractor-mounted or tractor-drawn field sprayers and aircraft. Special equipment is used for soil-applied granular formulations. In the Mediterranean region, where spraying is intense, tractor-operated field sprayers are normally used although, if crop growth or irrigation makes ground application difficult, aircraft are used. Aerial spraying may be done by the private sector, the cooperative unions or the Turkish Air Association.

In the Aegean region, where fields are small, there is more use of backpack sprayers or tractor-operated equipment than of aircraft. In the southeast Anatolia region, where the use of pesticides is still uncommon on cotton, tractor-operated sprayers and aircraft are used when necessary. Contamination of sensitive non-target areas, such as canals and pastures, is often unavoidable during spraying operations, especially aerial spraying.

RECOMMENDATIONS

Cotton production costs, especially the cost of pest control, are continually increasing and prices received for cotton do not keep pace. Consequently, farmers are increasingly turning to alternative crops. If a decline in cotton production is to be prevented a programme of research and development on ways of reducing pest control costs needs to be implemented. The main requirements of the programme may be summarized as follows:

* Integrated pest management (IPM) is the most economic and effective way of reducing losses to pests and production costs in cotton.

• Economic threshold levels (ETLs) should be the criteria for pesticide application. ETLs for all pests need to be established and kept under constant review to ensure that pesticides are not used unnecessarily.

• Pest forecasting systems should be established, especially for Lepidopterous pests. Pheromone and other types of traps have potential for determining the timing of sprays, particularly for bollworm and aphid.

• Awareness among farmers and technical personnel of the contribution made to cotton pest control by natural enemies needs to be enhanced through training courses and demonstrations.

• Target-specific pesticides are to be preferred to broad spectrum chemicals. Spray application decision criteria should include the status of natural enemies as well as pests. Delaying spraying to allow natural enemies to be effective should be practised where possible.

• Varietal resistance to pests and diseases should continue as an important objective of plant breeding programmes.

• New methods of control, including biological control, microbial control and the use of insect growth regulators and other new chemistry, need to be investigated.

Components of the IPM recommendations for cotton in Turkey

The following cultural techniques should be practised:
• sowing cotton on the ridge;
• use of increased inter- or intrarow spacing;
• early sowing;
• optimal irrigation practices;
• optimal use of fertilizer;
• avoidance of foliar-applied fertilizers.

Planting material should be:
• of certified seed only;
• an appropriate choice of variety for the location.

A crop rotation of wheat-second crop maize/soybean-cotton-wheat should be used.

Chemical control should adopt the following practices and principles:

- applications that are based on economic thresholds;
- applications that are related to crop phenology;
- rotation of pesticide groups to inhibit resistance development;
- avoidance of blanket applications where possible – use of strip or field border applications instead;
- use of selective pesticides;
- selection of the most appropriate application method.

Mechanical/manual control should concentrate on:
- attention to field sanitation;
- collection of rosetted flowers.

Regulations designed to keep pink bollworm from becoming a major pest should be observed.

KEY PESTS
Key pests of cotton in the Mediterranean region of Turkey are: whitefly *(Bemisia tabaci)*, American bollworm *(Helicoverpa armigera)*, aphid*(Aphis gossypii)*and red spider mite *(Tetranychus cinnabarinus)*.

In other regions the main pests are: aphid*(Aphis gossypii)* and two-spotted spider mite *(Tetranychus urticae)*.

KEY PERSONNEL INVOLVED IN COTTON PEST MANAGEMENT IN TURKEY
Cafer Mart Ph.D., National Coordinator in entomology at the Plant Protection Research Institute (PPRI), Adana, Turkey.

Saban Karaat Ph.D., entomologist at PPRI, Adana.

Erkin Ulug M.Sc., weed scientist at PPRI, Adana.

Atilla Atac M.Sc., plant pathologist at PPRI, Adana.

Izzet Kadioglu Ph.D., weed scientist at PPRI, Adana.

Ilhan Uremis M.Sc., weed scientist at PPRI, Adana.

Veli Cetin M.Sc., plant pathologist at PPRI, Adana.

Ayhan Karcilioglu Ph.D., plant pathologist at the Plant Protection Research Institute (PPRI), Bornova-Izmir, Turkey.

Fusun Tezcan Ph.D., entomologist at PPRI, Bornova-Izmir.

Abdurrahman Uzun Ph.D., weed scientist at PPRI, Bornova-Izmir.

Emin Onan Ph.D., plant pathologist at PPRI, Bornova-Izmir.
Abuzer Sagir Ph.D., plant pathologist at the Plant Protection Research
Institute (PPRI), Diyarbakir, Turkey.
M. Ali Goven Ph.D., entomologist at PPRI, Bornova-Izmir.
Ahmet Ulugad M.Sc., weed scientist at PPRI, Diyarbakir.
Fahri Tatli M.Sc., plant pathologist at PPRI, Diyarbakir.

References

Atakan, E. & Ozgur A.F. 1994. Research on efficiency of natural enemies on population development of cotton aphid, *Aphis gossypii* Glov. *Proceedings of the 3rd Turkish Conference of Plant Pests.*

Bakirci, H. 1970. Research on weeds and their control on cotton areas in Aegean region. Faculty of Agriculture, University of Ege, Bornova-Izmir. (M.Sc. thesis)

Belli, A., Tunc, A., Turhan, N., Yabas, M.N., Kismir, A., & Kisakurek, N. 1983. Preliminary studies on overwintering conditions and adult lifespan of the pink bollworm, *Pectinophora gossypiella* Saund. in the Adana region. *Bitki Koruma Bull.* 23:4, 207-222.

Cengiz, J. & Tezcan, F. 1987. Investigations on the role of some synthetic pyrethroids in increasing red spider mite populations, *Tetranychus urticae* (Koch.) and *T. cinnabarinus* Boisd., on cotton. *Proceedings of the 1st Turkish National Congress of Entomology,* p. 287-296. University of Ege, Bornova-Izmir, Turkey.

Cengiz, J., Yalcin, E. & Tezcan, F. 1987. Research on electrodyn with CDA (controlled droplet application) on red spider mites, *Tetranychus urticae* (Koch.) and *T. cinnabarinus* Boisd., on cotton. *Proceedings of the 1st Turkish National Congress of Entomology,* p. 335-339. University of Ege, Bornova-Izmir, Turkey.

Dincer, J., Karaman, M., Kavut, N., Zumreoglu, S., Cengiz, F., Pala, Y., Uysal, M. & Gonen, M. 1981. *Research on possibilities of integrated pest management against cotton pests in Aegean region.* Final Report of Project A 105 013. PPRI, Bornova-Izmir, Turkey.

Esentepe, M. 1979. Research on verticillium wilt factor and its distribution, density, degree of injuriousness and ecology in Adana and Antalya. *Gida, Tar. ve Hay. Bak., Zir. Muc ve Zir. Kar. Gn Md. Aras. Eser.,* No. 32.

Esentepe, M., Karcilioglu, A., Sezgin, E. and Onan, E. 1982. Research on the determination of the relationship between severe cotton verticillium wilt, *Verticillium dahliae* Kleb. and crop reduction in Aegean Region. *Proceedings of the 3rd Turkish National Congress of Phytopathology,* p. 145-151.

Gencer, O. 1991. Place of Turkish cotton production in the world and important problems of textiles, *Tekstil Maraton,* July, p. 17-25.

Ghavami, M.D. & Ozgur, A.F. 1992. Population development of pests and their interaction with predatory insects in cotton fields. *Proceedings of the 2nd Turkish National Congress of Entomology,* p. 227-238. Ege University, Bornova-Izmir, Turkey.

Isler, N. 1987. Research on effect on population development of cotton whitefly, *Bemisia tabaci* Genn., and cotton production, of different sowing time and method, irrigation method, fertilization and two sprays applied at different times. Plant Research Institute, Cukorova University, Adana, Turkey. (Ph.D. thesis)

Kadioglu, I., Ulug, E. & Uremis, I. 1993. Research on weeds in cotton areas in the Mediterranean region. *Proceedings of the 3rd Turkish National Congress of Plant Biology,* p. 151-156, Adana, Turkey.

Karaat, S. 1991. Biological parameters and population changes of *Tetranychus urticae* Koch. on candidate cotton varieties for southeast Anatolia. Plant Research Institute, Cukorova University. (Ph.D. thesis)

Karaat, S. & Goven, M.A. 1985. The first record of *Phoranous varicoloreus* Burm. (Coleoptera: Scarabaeidae: Rutelinae) as a harmful insect on cotton in southeast Anatolia. *Bitki Koruma Dergisi, Turk.,* 9(1): 45-52.

Karaat, S. & Goven, M.A. 1985. Investigations into the effects of predacious species of insects against two-spotted spider mites, *Tetranychus urticae* Koch., in cotton fields in southeast Anatolia. *Bitki Koruma Dergisi, Turk.,* 9(4): 239-245.

Karaat, S., Goven, M.A. and Mart, C. 1986. General situation of natural enemies in the cotton areas in southeast Anatolia. *Proceedings of the 1st Turkish National Congress of Plant Pests,* Adana, Turkey.

Karaat, S., Goven, M.A. & Mart, C. 1987. The relationship between pests in cotton fields of the southeast Anatolia region and the phenology of the crop. *Proceedings of the 1st Turkish National Congress of Entomology,* p. 189-198.

Karaat, S., Goven, M.A. and Mart, C. 1989. *Research on establishing Integrated Pest Management against cotton pests,* Final Report of Project 5/A 200.001. PPRI, Diyarbakir, Turkey.

Karaat, S., Goven, M.A. & Mart, C. 1989. General situation of *Campylomma* spp. in cotton areas in south-east Anatolia. *Proceedings of the National Symposium of Plant Pests,* Antalya, Turkey.

Karaat, S., Goven, M.A. & Mart, C. 1992. Integrated pest management for cotton pests in southeast Anatolia Project (G.A.P.) areas. *Proceedings of the 2nd Turkish National Congress of Entomology,* p. 183-191. Ege University, Bornova-Izmir, Turkey.

Karaca, I., Karcilioglu, A. & Ceylan, S. 1971. Wilt disease of cotton in the Ege region of Turkey. *J. Turkish Phytopath.,* 1(1): 4-11.

Karaca, I., Ceylan, S. & Karcilioglu, A. 1973 The importance of cottonseed in the spread of verticillium wilt. *J. Turkish Phytopath.,* 2: 30-31.

Kaygisiz, H. 1976. Research on description, biology, distribution, damage, host plants and control methods of cotton whitefly, *Bemisia tabaci* Genn., in the Mediterranean region. *Tarim ve Orman Bakanligi, Zirai Mucadele ve Zirai Karantina Genel Mudurlugu Yayinlari, Arastirma Eserleri,* No. 45 (58 S.).

Killi, F. 1993. Determination of some annual and perennial weed densities in cotton fields in Kahramanmaras. *Proceedings of the 3rd Turkish National Congress of Plant Biology,* p. 157-161. Adana, Turkey.

Kismir, A. 1983. Importance of biological control in cotton pest management in Turkey. *Symposium on the Integrated Pest Control for Cotton in the Near East.* p. 177-191. 5-9 September 1983. FAO/UNEP Near East Integrated Country Programme for the Development and Application of Integrated Pest Control in Cotton Growing. Rome, Italy, FAO.

Kismir, A. 1989. Research on biology and ecology of predatory insect *Anisochrysa carnea* (Steph.) (Neuroptera: Chrysopidae) and its effect on bollworm, *Heliothis armigera* Hb. (Lepidoptera: Noctuidae). *Tarim Orman ve Koyisleri Bakanligi Arastirma Yayinlari,* No. 71 (97 S.).

Kismir, A. & Sengonca, C. 1980. Research on the effects on the predator,

Anisochchrysa carnea (Steph.) (Neuroptera: Chrysopidae), of some chemicals used against cotton pests in Cukurova region. *Bitki Koruma Dergisi, Turk.,* 4(40): 243-250.

Kismir, A., Tunc, A., Turhan, N., Belli, A. & Pala, U. 1984. *Research on the possibilities for integrated pest management of cotton pests in Mediterranean region,* Final Report of Project A 103 617/1. PPRI, Adana, Turkey.

Madran, N. 1971. *Cotton in Turkey.* Adana, Turkey, Ministry of Agriculture and Rural Affairs, Research Institute. No. 27 (91 S.).

Onan, E. 1982. Research on the effect of chemical fertilizer on virulence of *Rhizoctania solani* Kuhn. in cotton. *Proceedings of the 3rd Turkish National Congress of Phytopathology.*

Onan, E. 1993. Effect of soil solarisation on the viability of *Verticillium dahliae* Kleb. microsclerotia in the Aegean region of Turkey. *J. Turkish Phytopath.,* 22(2-3): 85-93.

Onan, E., Karcilioglu, A. & Cimen, M. 1994. Effect of soil solarisation in controlling verticillium wilt of cotton in the Aegean region. *J. Turkish Phytopath.,* 23(1): 1-7.

Ozgur, A.F. & Isler, N. 1992. Effect of irrigation and fertilization on plant development, cotton production and population development of cotton whitefly, *Bemisia tabaci* Genn. *National Symposium of Integrated Pest Management,* p. 227-234, Bornova-Izmir, Turkey.

Ozgur, A.F., Sekeroglu, E., Gencer, O., Gocmen, H., Yelin, D. & Isler, N. 1988. Research to determine the relationship between the development of populations of cotton pests and crop phenology and variety. *Doga, Turk Tarim ve Ormancilik Dergisi,* 12(1): 48-74.

Pala, Y. 1990. Research on the relationship between pest life cycles, plant development and yield in two varieties of cotton. *Tarim Orman ve Koyisleri Bakanligi Arastirma Yayinlari,* No. 71 (97 S.).

Sengonca, C. & Yurdakul, O. 1975. The economic effects of the cotton whitefly, *Bemisia tabaci* Genn. epidemic in Cukurova Region. Cukurova University, *Agriculture Faculty Annals,* 2: 137-148.

Sezgin, E., Karcilioglu, A. & Esentepe, M. 1977. Research on the role of diseased cotton litter in the spread of *Verticillium dahliae* Kleb. *Plant Protection Research Bulletin,* 11: 85-86.

Sezgin, E., Karcilioglu, A. & Yemiscioglu, U. 1982. Investigations on the effects of some cultural applications and antagonistic fungi on *Rhizoctonia solani* Kuhn. and *Verticillium dahliae* Kleb. in the Aegean region. I. Effects of crop rotation and fertilizer. *J. Turkish Phytopath.*, 11(1-2): 41-54.

Sezgin, E., Karcilioglu, A. & Yemiscioglu, U. 1982. Investigations on the effects of some cultural applications and antagonistic fungi on *Rhizoctonia solani* Kuhn. and *Verticillium dahliae* Kleb. in the Aegean region. II. Effects of herbicides and antagonistic fungi. *J. Turkish Phytopath.*, 11:(3), 79-91.

Stam, P.A. & Tunc, A. 1983. Importance and need for Integrated Pest Management in Turkey. *Symposium on Integrated Pest Control for Cotton in the Near East.* p. 145-153. 5-9 September 1983. FAO/UNEP Near East Integrated Country Programme for the Development and Applications of Integrated Pest Control in Cotton Growing, Rome, Italy, Rome.

Stam, P.A. & Tunc, A. 1984. *Development and application of integrated pest control on cotton. Prospects of integrated control on cotton in the Cukurova region of Turkey.* AG:DP/TUR/83/008, Field Document No. 1. Rome, Italy, FAO.

Tezcan, F. 1991. Research on factors affecting the biology and population dynamics of the cotton aphid, *Aphis gossypi* Glov. (Homoptera: Aphididae), on cotton in Izmir and Manisa. Faculty of Agriculture, University of Ege, Bornova-Izmir, Turkey. (Ph.D. thesis)

Tezcan, F. & Ozmen, B. 1992. Some biological studies on the cotton aphid, *Aphis gossypi* Glov. (Homoptera: Aphididae), under laboratory conditions. *Proceedings of the 2nd Turkish National Congress of Entomology*, p. 417-423. Ege University, Bornova-Izmir, Turkey.

Tunc, A., Turhan, N., Belli, A., Yabas, M.N., Kismir, A., Tekin, T. & Kisakurek, N. 1980. *Research on the possibilities for integrated pest management on cotton in the Mediterranean region.* Final Report of Project A 103617/1. PPRI, Adana, Turkey.

Tunc, A., Turhan, N., Belli, A., Kismir, A., Tekin, T. & Kisakurek, N. 1983. Investigations to determine the winter survival and winter host plants of the whitefly *Bemisia tabaci* Genn., in the Cukurova Region. *Bitki Koruma Bult.*, 23:(1), 42-51.

Ulugad, A. & Katkat, M. 1991. Research to determine the distribution and density of weeds in cotton areas in southeast Anatolia region. *Proceedings of the 4th Turkish National Congress of Phyotopathology,* p. 125-131, Bornova-Izmir, Turkey.

Uygur, F.N., Koch, W. & Walter H. 1986. *Descriptions of some important weeds in the cotton - wheat rotation in the Cukurova region.* PLITS/4.1. Hohenheim University, Stuttgart, Germany.

Van Gent, R. 1984. *Development and application of integrated pest control on cotton in Turkey. Preliminary observations on parasitization of whitefly,* Bemisia tabaci *Genn., on cotton in the Cukurova region, Turkey.* Ag:DP/TUR/83/008, Field Document No. 2. Rome, Italy, FAO.

Yabas, M.N. 1983. Importance and management of *Heliothis* in cotton growing in Turkey. *Symposium on Integrated Pest Control for Cotton in the Near East,* p. 79-90. 5-9 September 1983. FAO/UNEP Near East Integrated Country Programme; Integrated Pest Control in Cotton Growing. Rome, Italy, FAO.

Zuhal, I. 1972. Research on weed species and their life cycles, distribution, and damage in irrigated and non-irrigated cotton in Menemen. Faculty of Agriculture, University of Ege, Bornova-Izmir, Turkey. (Ph.D. thesis)

Annex

SUBPROJECTS OF NATIONAL INTEGRATED PEST MANAGEMENT PROJECT FOR COTTON IN TURKEY

Name of project	Purpose	Principal investigator
1. Studies on the development of control methods of key insects on cotton in Aegean region	Development of economic injury thresholds for cotton aphid, *Aphis gossypii* Glov.	Dr F. Tezcan (PPRI, Bornova-Izmir)
2. Studies on chemical control of cotton whitefly, *Bemisia tabaci* Genn., in Mediterranean region	a) Determination of the status of cotton whitefly b) Determination of biological activities of insecticides used against cotton whitefly	Dr C. Mart (PPRI, Adana)
3. Studies on establishment of forecasting and warning systems for cotton pests in Mediterranean region	Determination of critical treatment time using light and pheromone traps for bollworm	Dr C. Mart
4. Studies on the effect of pest population density on some morphological and physiological characters of cotton	Determination of the effect of population density of *Aphis gossypii, Empoasca* spp. and *Tetranychus* spp. on some morphological and physiological characters of cotton	H. Dundar (CRI, Nazilli)
5. Studies on characters conferring resistance to *Aphis gossypii* Glov. in cotton varieties in Mediterranean region	Determination of aphid populations on cotton varieties that have different physiological and morphological characters	Dr C. Mart
6. Studies on the susceptibility of cotton varieties to *Verticillium dahliae* Kleb.	Determination of reactions to disease factors of some cotton varieties selected for the southeast region of Turkey	Dr A. Sadir (PPRI, Diyarbakir)
7. Studies on the effects of chemical fertilizer on cotton damping-off in the field	Determination of the effects of fertilizer application time and dosage on cotton damping-off in cotton fields	M. Cymen (PPRI, Bornova-Izmir)
8. Studies on possible control methods for *Verticillium dahliae* Kleb. in Aegean region	Determination of the effects of soil-solarization on the microsclerotia of *V. dahliae*	Dr E. Onan (PPRI, Bornova-Izmir)
9. Studies on the susceptibility of cotton varieties to *Verticillium dahliae* Kleb. in Cukorova region	Determination of susceptibility of some varieties to *V. dahliae* in Mediterranean region	A. Atac (PPRI, Adana)
10. Studies on control methods of *Sorghum halepense* (L.) Pers. on cotton fields in southeast Anatolia region	Determination of the effect of combining different tillage techniques and herbicides and the optimum treatment time against *S. halepense* in cotton	A. Uludag (PPRI, Diyarkabir)
11. Studies on the effect of *Sorghum halepense* (L.) Pers. on cotton fields in southeast Anatolia region	Determination of the effect of different *S. halepense* densities on cotton production	A. Uludag
12. Studies on weeds in cotton fields in Mediterranean region	Determination of weed species and density and their effects on yield of cotton	Dr I. Kadioglu (PPRI, Adana)

Yemen

Saeed A. Ba-Angood

INTRODUCTION

History

The history of cotton in Yemen started when the Sudanese-Egyptian variety X1730A, a Lambert type of long-staple cotton (*Gossypium barbadense*) was imported from the Sudan in 1946 and released for commercial production in the Abyan Delta area in August 1950. An area of some 4 000 ha was sown, giving an average yield of 650 kg of seed cotton per hectare, with 80 percent in grades 1 to 3 (Muallem, 1968).

Since 1946 the X1730A variety has been reselected to produce a succession of new releases. In 1955, AB-1 was released, followed in 1957 by AB-3. A variety screening programme led to the release in the 1959/60 season of Bar XL1, a bacterial blight resistant variety of Sudanese origin. Selections of this variety were released in subsequent years, K1 in 1960/61, followed by K4. A reselection of K4 was released in 1966/67 and now occupies the whole of the long-staple production in eastern Yemen (Muallem, 1968; Ba-Angood, 1991). Crosses between K4 and Giza 68 have produced a number of promising new strains, including KB-138, KB-226 and KB-277, which may be released for commercial cultivation in future years. The whole of the long-staple production is exported and is well known in the international market as "Abyan cotton" (Muallem, 1985).

The evaluation of medium-staple, American upland (*G. hirsutum*) varieties began on a small scale in the 1959/60 season. In 1970 a programme of local adaptability trials of a large number of imported medium-staple varieties began. Varieties tested included Coker 100 Wilt, Coker 201, Coker 310, Acala 4-42, Acala 1517B, Acala Sj1 and Acala Sj2 (Muallem, 1985). Commercial production of medium-staple varieties began in the period 1974 to 1976 when Coker 100 Wilt was released in the Mudia, Ahwar and Maifaa Hajar areas of the eastern governorates.

In the northern areas of Yemen commercial cotton growing began in the 1951/52 season in the Tihama area where 118 ha were sown, producing 132 tonnes of seed cotton. The area under cultivation reached a peak of 29 254 ha in the 1974/75 season when production of seed cotton was 27 174 tonnes (Al-Gifri, 1984). Both long-staple Sakel types and medium-staple varieties, including Acala 1517B and Coker 310, were introduced in this area. Since the 1986/87 season Acala Sj2 has been the recommended variety.

Production

The main production areas are now the coastal governorates of Abyan (Abyan Delta – long-staple, K4) and Lahej (Tuban Delta – medium-staple, Coker 100 Wilt) and Al-Hodeidah governorate in northern Yemen (Tihama – medium-staple, Acala Sj2). A few hectares are also grown in Mareb, Dhamar and Ibb governorates (Table 36).

Table 37 gives production data for selected seasons between 1973/74 and 1992/93. The considerable year-to-year variation in the area grown is partly attributable to the facts that cotton is dependent on spate irrigation and that floods vary both in timing and quality.

Socio-economic factors in cotton production

When cotton production began in the eastern areas of the southern part of Yemen it received support from the government which established the cotton boards for the Abyan and Tuban deltas. These boards gave interest-free loans and provided other inputs for cotton growers, who were organized in cooperatives or on state farms. Cotton pest control was carried out by the Ministry of Agriculture and Agrarian Reform (MAAR). Farmers were content with cotton in the 1960s, but as time went on production costs rose at a faster rate than government prices for cotton and higher returns could be obtained from other crops such as vegetables and forage, so cotton production declined. It received a temporary boost when the government raised prices in 1987, but subsequently declined again. For example, Abyan production was 15 212 tonnes of seed cotton in 1965/66 but only 1 000 tonnes in 1989/90 (Ba-Angood, 1990).

In the northern areas cotton production began in 1951/52 and the area

TABLE 36
Cotton area and production in Yemen by governate, 1989/90

Governorate	Area *(ha)*	Production *(tonnes seed cotton)*
Al-Hodeidah	7 647	6 147
Abyan	1 963	1 014
Lahej	651	594
Mareb	68	54
Dhamar	27	24
Ibb	7	6
Total	**10 363**	**7 839**

Source: Ministry of Agriculture and Water Resources (MAWR), 1991, *Agricultural Statistics for 1990.*

TABLE 37
Cotton area, production and yield

Season	Area *(ha)*	Production *(tonnes seed cotton)*	Yield *(kg seed cotton per ha)*
1973/74	39 080	36 500	934
1982/83	12 080	10 970	908
1983/84	8 740	7 010	802
1984/85	8 450	7 000	828
1987/88	9 642	7 810	810
1988/89	15 782	13 036	826
1989/90	10 363	7 839	756
1990/91	9 445	7 291	772
1991/92	18 683	12 622	676
1992/93	14 103	10 762	762

Source: Arab Organization for Agricultural Development (AOAD), 1987, *Yearbook of Agricultural Statistics* vol. 7, December 1987; Ministry of Agriculture and Water Resources (MAWR), 1993, *Agricultural Statistics Yearbook 1992.*

under cotton steadily increased so that by 1974/75 it had reached 29 254 ha. Subsequently it declined rapidly, to only 2 593 ha in 1979/80. The reasons for this decline included shortages of labour as people left to work in Saudi Arabia, Qatar, Bahrain and the United Arab Emirates, high production costs and low prices for cotton and insect problems. To stem the decline in production the government formed the Cotton General Company charged with providing cotton growers with credit, purchasing and ginning seed cotton, supplying lint to local textile mills and exporting surplus production. The Ministry of Agriculture and Fisheries (MAF) became responsible for pest control. As a result the area under cotton increased to 7 000 ha in 1981/82 and production increased as well, from 3 700 tonnes of seed cotton in 1983/84 to 5 600 tonnes in 1987/88 (Ba-Angood, 1990), but never regained the level reached in 1974/75.

In 1990, following unification, the various cotton boards and the Cotton General Company were no longer able to service cotton growers because of financial problems, and production declined. Infrastructural problems are now being addressed by the authorities, including the Ministry of Agriculture and Water Resources (MAWR), the Tihama Development Authority (TDA) and the various cotton boards, with the intention of reviving the cotton industry.

CULTURAL PRACTICES
Land preparation and crop management
Cotton is grown under spate irrigation. Following the arrival of the first floods, fields are heavily irrigated and then left for a few days until the soil is in a workable condition. Fields are then deep-ploughed using disc-ploughs and disc-harrowed to prepare the seed bed.

Long-staple cotton is sown between 15 August and 15 September and is harvested between January and the end of April. Cotton may remain in the field until May or June. Medium-staple cotton is sown earlier, between 15 July and 30 August, and is harvested between December and early March.

Seed rates vary from 14-16 kg per hectare for long-staple varieties, with spacings of 70 cm between rows and 40 cm between plants, to 15-20 kg per hectare for medium-staple varieties.

Fertilizer

In eastern Yemen floodwaters cover the fields in silt that is rich in calcium and phosphorus so the application of phosphatic fertilizer (P_2O_5) is not necessary. Nitrogen (N) is applied as urea (46 percent N) at a rate of 50 kg per hectare, usually at the time of land preparation (Atta, 1984). In northern Yemen recommendations for cotton grown on alluvial soils under spate irrigation or under spate supplemented with pump irrigation, suggest that N is applied at the rate of 100 kg per hectare, plus 50 kg P_2O_5 per hectare (Al-Gifri, 1984).

Irrigation

Most cotton is irrigated from spates arising as a result of rainfall in the mountainous areas. These floods are directed on to fields by systems of weirs and canals (Muallem, 1968). Most cotton therefore receives only one irrigation, before sowing, and is grown on residual moisture. However a few areas, mainly in northern Yemen, receive supplementary groundwater and rain-fed irrigations.

There have been a number of investigations into the water requirements of cotton in Yemen (Ogborn, 1960; Anthony, Ogborn and Proctor, 1961; Farbrother, 1962; Rijks, 1965; Al-Shubaihi, 1978). The results have been conflicting, with regard both to the amount of water required to grow cotton and to the effects of water stress on cotton growth (Atta, 1984). Rates for water application for economic cotton production have been variously suggested as 45 cm^3 (Al-Shubaihi, 1978); 60 cm^3 (Muallem, 1982); and 85 cm^3 (Rijks, 1965). In northern Yemen Al-Gifri (1984) reported that water requirements for cotton were 80 cm^3 from eight or nine irrigations at 14-day intervals, with a contribution from rainfall of 20 cm^3.

PESTS
Insects and mites

The main insect and mite species recorded on cotton in Yemen are shown in Table 38, which shows their relative economic importance in the northern and eastern cotton-growing areas. Bollworms, especially red bollworm, are of major importance in both areas, with the American and spiny bollworms

TABLE 38

Insect and mite pests recorded on cotton in Yemen

Pest common name	Scientific name	Status	
		Northern areas	Southern areas
Termites	*Microtermes najdensis* Harris	+++	
	Microcerotermes diversus Silvestri	+	+
Red bollworm	*Diparopsis watersi* (Roths.)	+++	+++
Spiny bollworm	*Earias insulana* (Boisd.)	+++	++
American bollworm	*Helicoverpa armigera* (Hb.)	+++	++
Pink bollworm	*Pectinophora gossypiella* (Saund.)	+	
Whitefly	*Bemisia tabaci* (Gen.)	+++	+
Aphid	*Aphis gossypii* (Glov.)	+++	++
Jassid	*Jacobiasca lybica* (De Bergevin)	+++	+
Leaf worm	*Spodoptera littoralis* (Boisd.)	++	+
Lesser armyworm	*S. exigua* (Hb.)	+	+
Leafroller	*Syllepte derogata* (Fabricius)	++	+
Semi-looper	*Xanthodes graellsii* (Feisth.)	++	+
Green stink bug	*Nezara viridula* (L.)	++	
	Acrosternum millieri (Mulsant & Reg) (Het.: Pentatomidae)	+	
Semi-looper	*Anomis flava* (F.)	++	
Cottonseed bug	*Oxycarenus hyalinipennis* (Costa)		++
Flea beetle	*Podacgrica puncticollis* Weise.	+	
	P. fuscicornis (L.)	+	
Cotton stem borer	*Sphenoptera gossypii* Cotes (Col.: Buprestidae)		++
Blue bug	*Calidea dregii* Germar (Het.: Scutellaridae)	+	
	Hamartus instabilis Marshall (Col.: Curculionidae)	+	
	Dicranocephalus sp. (Het.: Stenocephalidae)	+	
Cotton stainer	*Dysdercus* sp.	+	
	Grammodes stolida (F.) (Lep.: Noctuidae)	+	
Chafer grub	*Schizonycha* sp. (Col.: Scarabaeidae)	+	
	Spilostethus pandurus militaris (F.) (Het.: Lygaeidae)	+	
	Spilostethus sp.	+	
	Stalagmosoma sp. (Col.: Scarabaeidae)	+	
Lygus	*Taylorilygus vosseleri* Popp	+	
Cotton thrips	*Thrips tabaci* (Lindeman)	+	+
	Frankliniella schultzei (Trybom)	+	+
	Utetheisa pulchella (L.) (Lep.: Arctiidae)	+	+
	Julodis sp. (Col.: Buprestidae)	+	
	Longitarsus sp. (Col.: Chrysomelidae)		
Blister beetle	*Mylabris maculiventris* Klug. (Col.: Meloidae)	+	+
	Mylabris sp.	+	
Cotton mealybug	*Ferrisia virgata* (Cockerell) (Hom.: Pseudococcidae)		+
Oriental mite	*Eutetranychus orientalis* (Klein)	+	+
Red cotton mite	*Tetranychus cinnabarinus* (Boisd.)	+	+

Source: Nasseh and Mahyoub, 1987; Ba-Angood, 1991.
+++ = very common;
++ = common;
+ = infrequent.

being of particular importance in the northern areas. In these areas termites are also a major problem. Sucking pests, including whitefly, aphid and jassid, are also important, especially in the northern areas.

Insects are more important causes of yield loss than either diseases or weeds in all the cotton-growing areas of Yemen. A number of authors, including Anthony, Ogborn and Proctor (1961), Ba-Angood (1972), Zaazou *et al.* (1976), Al-Ghashm (1985); Ba-Angood, Abdessattar and Masoud (1985) and Ba-Angood (1991), have discussed the factors, including pest attack, that affect cotton yields. In the Abyan and Lahej deltas the bollworm complex, in particular the red bollworm, can cause 10 to 30 percent crop loss and in some seasons this may reach 50 percent. In 1975 pink bollworm caused 50 percent damage at the El-Kod cooperative, but since then it has been of no economic importance to the eastern areas (Ba-Angood, 1991).

In the northern, Tihama, cotton-growing areas termites, especially the najd termite, can reduce stands by 60 to 100 percent during the first six weeks after sowing if seed is not treated with aldrin (Al-Zabeidi, 1985).

The cottonseed bug has recently emerged as a late-season pest of some importance in the eastern cotton-growing areas. It causes loss of lint quality through staining and also reduces the oil content and viability of cottonseed. No control measures are recommended. The red spider mite and the oriental mite are both becoming more important as pests, although there is considerable variation between locations and from season to season in the status of these mites. The cotton stem borer, combined with Abyan root rot, is a major cause of stand loss in the Abyan Delta in the first six weeks after sowing.

Diseases

Table 39 shows the main diseases of cotton recorded in Yemen and their importance in the main cotton-growing areas. In the eastern areas Abyan root rot, a fungal disease of the roots, causes wilting of young plants and, together with the cotton stem borer, can cause 10 to 20 percent damage. Root rot usually occurs in patches, but sometimes individual plants are observed to be affected up to two months after sowing (Ba-Angood, 1991). Other main diseases of the eastern area include angular and alternaria leaf spots and boll rots. In the northern areas damping-off disease and angular leaf spot are the main disease problems.

TABLE 39

Cotton diseases recorded in Yemen

est common name	Scientific name	Status	
		Northern areas	Southern areas
Abyan root rot	*Rhizoctonia solani*Kuhn	+	++
Root rot	*Thanatephorus cucumeris*(Frank.) Donk.	++	
Boll rot	*Aspergillus flavus*Link.	++	++
Fusarium wilt	*Fusarium oxysporum*Schlecht f. sp. *vasinfectum* Alk. Sny. & Hans	++	+
Bacterial blight	*Xanthomonas campestris pv malvacearum* (E.F. Smith) Dye	+	+
Leaf spots	*Alternaria gossypina* (Thum) Hopkins	+	+
	A. macrospora Zimm	+	+
	*Cercospora gossypina*Cooke	+	
Verticillium wilt	*Verticillium dahliae*(Kleb.)		+

Source: Kamal and Aghbari, 1985; Ba-Angood, 1991.
++ = moderate infection;
+ = negligible.

Weeds

Table 40 shows the weed species recorded in cotton in Yemen. The most important genera are *Cynodon, Cyperus, Echinochloa, Solanum* and *Heliotropium* (Kassasian, 1980; Ba-Angood, Abdessattar and Masoud, 1985). There are no data on yield losses in cotton to weed competition in Yemen but the weed problem is considered to be most serious in young cotton.

Nematodes

Nematodes are not economic pests of cotton in Yemen.

CONTROL MEASURES
Chemical control

Table 41 lists the main pesticides recommended for use on cotton in Yemen. Monocrotophos and carbaryl have been recommended for bollworm control in the eastern cotton-growing areas, although monocrotophos is now being replaced by a mixture of fenvalerate and fenitrothion (Ba-Angood, 1991), especially in the Tuban Delta. Dimethoate is used for sucking pest control.

TABLE 40
Weeds recorded in cotton fields in Yemen

Family	Scientific name	Status
Amaranthaceae	*Aerva javanica* (Burm. f.) Juss.	+
	Digera muricata (L.) Mart.	+
	Amaranthus graecizans L.	+
Boraginaceae	*Heliotropium europaeum* L.	+++
	H. lasiocarpum C.A. Mey.	+
Capparidaceae	*Dipterygium glaucum* Decn.	+
Compositae	*Pulicaria undulata*	+
	Sonchus oleraceus L.	+
	Elipta prostrata (L.) L.	+
Convolvulaceae	*Convolvulus arvensis* L.	+
Cruciferae	*Capsella bursa-pastoris* (l.) Medik.	+
Cyperaceae	*Cyperus rotundus* L.	+++
Euphorbiaceae	*Chrozophora* sp.	+
	Euphorbia sp.	++
	E. hypericifolia L.	++
	Phyllanthus rotundifolia	+
Graminae	*Cynodon dactylon* (L.) Pers.	+++
	Dactyloctenium aegyptium (L.) Pers.	+
	Echinochloa colona (L.) Link	+++
	Pennisetum sp.	+
	Desmostachia bipinnata	++
	Chloris barbata Sw.	+
	Brachiaria reptans (L.) Gard. & Hubb.	+
	Phragmites australis (Cav.) Steud.	+
		+
Leguminosae	*Alhagi maurorum*	+
	Cassia italica (Mill.) Lam. ex Steud.	+
	C. occidentalis L.	+
	Crotalaria aegyptiaca	+
	Rhynchosia minima (L.) DC.	++
	Tephrosia appolinea (Del.) Link	+
		++
Malvaceae	*Abutilon pannosum* (Forst. f.) Schlecht	+
		+
Molluginaceae	*Glinus lotoides* L.	+
		+
Papaveraceae	*Argemone mexicana* L.	+++
		+
Plantaginaceae	*Plantago major* L.	+
		+
Portulacaceae	*Portulaca oleracea* L.	
Solanaceae	*Datura innoxia* Mill.	
	Solanum dubium Fres.	
	Withania somnifera (L.) Dun.	
Tiliaceae	*Corchorus depressus* (L.) Stocks	
	C. trilocularis L.	

Source: Kassasian, 1980; Ba-Angood, Abdessattar and Masoud, 1985.
+++ = very common; ++ = common; + = infrequent.

TABLE 41
Pesticides used for cotton pest control in Yemen

Common name and formulation	Pests controlled
Insecticides	
Aldrin 40% ds	Termite seed dressing
Carbaryl 85% ds	Bollworm, flea beetles
Cypermethrin 10% ec	Bollworm, leafworm
Deltamethrin 2.5% ec	Bollworm, aphid
Dimethoate 40% ec	Whitefly, jassid, aphid, thrips
Fenitrothion 50% ec	Whitefly, jassid
Fenpropathrin 10% ec	Whitefly, mites
Fenvalerate 20% ec	Bollworm
Malathion 50% ec	Whitefly, aphid
Monocrotophos 40% ec	Bollworm, aphid
Permethrin 25% ec	Bollworm, leafworm
Pirimicarb 50% wp	Aphid
Trichlorfon 80% wp	Bollworm, leafworm
Dimethoate 30% + fenvalerate 10%	Bollworm, thrips, mites
Fenvalerate 5% + fenitrothion 25%	Bollworm, aphid, thrips
Fungicide	
Carboxin	Seed dressing

Source: Al-Ghashm, 1990; Ba-Angood, 1991.

Aldrin and chlordane were recommended for termite control in the Tihama area, but are now being phased out. Neither is now used in the eastern areas. Alternatives to aldrin have been recommended (Ba-Angood, 1994), including diazinon, fenvalerate and chlorpyrifos, but some farmers consider these to be less effective.

Seedling damping-off disease is controlled by dressing seed with a mixture of carboxin 37.5 percent plus captan 37.5 percent applied at the rate of 300 g per 100 kg seed.

There is no commercial use of herbicides in cotton, although they are being investigated in trials at the El-Kod Agricultural Research Centre and the Faculty of Agriculture at the University of Aden.

Farmers normally rely on the Plant Protection Department of the Ministry of Agriculture and Water Resources (MAWR) or the Tihama Development Authority (TDA) to spray their cotton, using motorized backpack sprayers.

The average number of sprays ranges from one to four, depending on the severity of pest attack. Some farmers have their own backpack sprayers for use in emergencies. Seed dressings are applied by the General Cotton Company or by seed multiplication schemes.

Pesticides generally continue to give good results in cotton pest control and there are few problems with side-effects. In Tuban Delta some farmers find that monocrotophos encourages excessive vegetative growth at the expense of the yield. Late applications of carbaryl have occasionally been observed to result in outbreaks of the cotton mealybug (Ba-Angood, 1991) and reductions in the numbers of beneficial species were noted (Ba-Angood, 1990). There is some concern that the application of sprays late in the season could leave residues that are harmful to people harvesting cotton and to stock grazing the crop, as well as affecting lint quality.

Legislative control

Plant quarantine regulations are designed to keep exotic cotton pests and diseases out of Yemen, and the import and distribution of cottonseed is controlled by the government, as is the export of cotton lint. The movement of cotton seedlings and stalks from one area to another in the eastern areas is prohibited and cotton has to be uprooted and burnt by the end of May. From then until the third week of July a legally enforceable close season exists during which no cotton may be in the ground. The main object is to prevent the carryover of red bollworm. Over the past five years, however, there has been a decline in the observance of the close season.

Cultural control

Deep-ploughing after harvest is recommended to destroy red bollworm pupae in the soil. Deep-ploughing to a depth of 30 to 40 cm with a mould-board plough followed by post-sowing hoeing also gives control of *Cynodon dactylon*. Ba-Angood (1982a) has suggested that overwintering diapause pupae could be induced to produce an early adult emergence by flooding the previous season's cotton fields, which would lower soil temperature and, in turn, break diapause. The emerging adults would find no cotton, their only host plant, on which to lay eggs, so the carryover population to the next season would be reduced.

Pink bollworm larvae in seed cotton can be killed by spreading cotton out in the sun to raise its temperatures to the lethal level of 55 to 57°C; this control method is practised in the Abyan Delta. The cottonseed bug can also be induced to leave seed cotton after harvest by spreading the crop out in the sun.

Biological control
Ba-Angood (1982b) reported the results of a survey of natural enemies of bollworms in the Abyan Delta and concluded that they contributed to early-season control. The main predators of bollworms were the lace wing *Chrysoperla carnea* (Stephens) (Neuroptera: Chrysopidae) and *Cosmolestus pictus* (Klug.) (Hemiptera: Reduviidae). Major parasitoids included the tachinids *Sturmia imberbis* (Wied.) and *S. inconspicua* (Mg.) (Diptera: Tachinidae). Insecticides used for bollworm control reduce populations of these beneficials (Ba-Angood, 1990).

Resistant varieties
Resistance to insect pests is not found in the commercial varieties grown in Yemen.

INTEGRATED PEST MANAGEMENT
Ba-Angood (1982b) reported the implementation of a package of recommendations for bollworm control, which was termed a "chemico-cultural control programme". The package comprised the following measures:
- uprooting and burning cotton stalks after harvest;
- enforcement of the close season (end of May to third week of July);
- early sowing of cotton, between the third week of July and mid-September;
- use of economic thresholds to determine the need for insecticide applications (for bollworms, 18 to 20 percent of fruiting points infested);
- application of water to the previous season's cotton field to encourage early emergence of the diapause red bollworm population.

The implementation of this programme at Giar, Al Husin and Batais agricultural cooperatives reduced the number of spray applications needed from an average of more than four to two or less. Bollworm populations were reduced and yields increased (Ba-Angood, 1982a; 1982b; 1991).

INFRASTRUCTURAL SUPPORT FOR COTTON IPM
Government support
Research institutes. Following unification, all research centres came under the Agricultural Research and Extension Authority at Dhamar. In future research will come under the Ministry of Higher Education and Research. The research institutes currently working on cotton pest management are:
  • Department of Plant Protection, Faculty of Agriculture, University of Aden;
  • Department of Plant Production and Protection, Faculty of Agriculture, University of Sana'a;
  • El-Kod Agricultural Research Centre;
  • Agricultural Research and Extension Authority (AREA), Dhamar;
  • Surdod Experimental Station (Tihama Regional Research Station);
  • Agricultural Research Service.

Plant protection services. MAWR is responsible for the General Administration of Plant Protection which has branches in most of the governorates. The Regional Plant Protection Office in Aden is responsible for plant protection services in the southern and eastern governorates and provides cotton pest control services free to farmers. In the Tihama region the Tihama Development Authority (TDA) carries out cotton pest control operations.

Extension services. The Department of Research and Extension at Aden provides extension services to the southern and eastern governorates, while in the Tihama area TDA provides extension units. In future all extension units may be coordinated by the Agricultural Research and Extension Authority (AREA).

Private agencies
Apart from the pesticide companies that import and distribute agrochemicals there is no private-sector involvement in cotton IPM.

Infrastructural problems
IPM in Yemen suffers from the following problems:
- shortage of trained staff;
- lack of research and extension capacity;
- poor dissemination of information;
- poor coordination and cooperation among research, extension and plant protection agencies;
- decline in farmers' interest in cotton which reduces pressure for change on the infrastructure;
- before unification there was little cooperation between northern and southern agencies and promotion of a unified approach is taking time to have an effect;
- free services provided by the cotton boards and the General Cotton Company can no longer be relied on and farmers are not receiving payment for their cotton on time.

GOVERNMENT AND FOREIGN ASSISTANCE IN COTTON IPM
Previous governments in the south of Yemen included plant protection projects in the three-year plans and in the first five-year plan. These projects were concerned with the establishment of plant protection centres in the various governorates. At El-Kod a programme of pesticide screening for bollworm control in IPM systems was carried out between 1970 and 1980. At Surdod termite control was investigated. Nasir's College of Agriculture is currently testing pesticides and developing IPM packages for the Lahej governorate. There are at present no foreign assistance projects specifically for cotton pest control, but various FAO/United Nations Development Programme (UNDP) and Yemen-German projects cover plant protection generally.

RECOMMENDATIONS
The infrastructure supporting cotton IPM, including research, plant protection and extension services, requires strengthening through the provision of equipment and facilities and staff training.

The components of IPM systems in Yemen require further research and development to make them more location-specific.

Pest forecasting and monitoring systems need further research and development and economic threshold levels need to be established.

Varietal resistance to pests and diseases should be an objective of cotton breeding programmes.

The package of chemico-cultural measures for bollworm control in Abyan Delta should be reintroduced.

The government needs to take measures to encourage cotton production, including reviewing cotton prices to make wider differentials between prices for different grades of cotton, the provision of credit for cotton farmers and the strengthening of research and extension.

Cotton's potential contribution to many sectors of the national economy should be recognized and investment made accordingly.

Cotton requires an overall, official, coordinating organization.

Regional cooperation in cotton production, plant production, processing and marketing is needed, possibly under the auspices of FAO or the Organization of Petrol Exporting Countries (OPEC).

A regional IPM project is needed to stimulate the adoption and development of this approach to cotton pest management to ensure sustainable production.

KEY PESTS OF COTTON IN YEMEN

The main pests of cotton in Yemen are: red bollworm (*Diparopsis watersi*), spiny bollworm (*Earias insulana*), American bollworm (*Helicoverpa armigera*), termite (*Microtermes najdensis*), aphid (*Aphis gossypii*), jassid (*Empoasca lybica*) and whitefly (*Bemisia tabaci*).

KEY PERSONNEL INVOLVED IN COTTON PEST MANAGEMENT IN YEMEN

Prof. Dr Saeed A. Ba-Angood, specialist in integrated pest management, Nasir's College of Agriculture, University of Aden, Khormaksar, P.O.Box 6172, Aden, Yemen.

Dr Haj S. Ba-Hamish, plant pathologist, Agricultural Research and Extension, Dhamar, Yemen.

Mr Sami. G. Hamshari, Director and specialist in plant protection, Regional Centre of Plant Protection, MAWR, Khormaksar, Aden, Yemen.

Dr M.Y. Al-Ghashm, Director and specialist in entomology, General
Administration of Plant Protection, MAWR, Sana'a, Yemen.

Dr Amin Al-Himyari, entomologist, Department of Plant Production and
Protection, Faculty of Agriculture, University of Sana'a, Sana'a, Yemen.

Mr Saeed A. Mahfoodh, entomologist, El-Kod Agricultural Research
Centre, MAWR, El-Kod, Abyan Governorate, Yemen.

Dr Abbass Ba-Wazir, specialist in weed control, Department of
Agronomy, Nasir's College of Agriculture, University of Aden, Al-Hawtah,
Lahej Governorate, Yemen.

References

Al-Ghashm, M.Y. 1985. *Cotton pests in YAR.* A country report presented at a
consultancy meeting on integrated pest management and rationalization of use
of chemical pesticides, 16 to 20 September 1984, Algiers, Algeria. p. 170-180.

Al-Ghashm, M.Y. 1990. *A pesticide manual for Republic of Yemen,* 2nd edition.
Agricultural Research and Extension Authority, Ministry of Agriculture and
Water Resources, Al-Mustagbal Press.

Al-Gifri, G.H. 1984. *Cotton in Yemen Arab Republic.* Tihama Development
Authority, Ministry of Agriculture and Fisheries, Sana'a, Yemen.

Al-Shubaihi, H. 1978. *A study on water requirement of the medium-staple cotton.*
El-Kod Agricultural Research Centre, Ministry of Agriculture and Agrarian
Reform, Aden, Yemen. (In Arabic)

Al-Zabeidi, A.H. 1985. *Methods of controlling termites and evaluation of losses
and damage on cotton.* 1st National Conference on Plant Protection in Yemen.
12 to 16 October 1985, Department of Plant Protection, Ministry of Agriculture
and Fisheries, Sana'a, Yemen. (In Arabic)

Anthony, K.R.M., Ogborn, J.E. & Proctor, J.H. 1961. Factors affecting cotton
yield in the Aden Protectorate. *Cotton Grow. Rev.,* 38: 161-71.

AOAD. 1987. *Yearbook of Agricultural Statistics* Vol. 7. Khartoum, the Sudan,
Arab Organization for Agricultural Development.

Atta, S. 1984. *Improvement of cotton in PDRY.* A country report submitted for
international workshop/training course on cotton production under heat and

drought stresses, 4 to 9 November 1984, Karachi, Pakistan. Pakistan Central Cotton Committee/FAO.

Ba-Angood, S.A.S. 1972. *Annual Report on Cotton Pests in PDRY.* Plant Protection Section, Agricultural Research Centre, Ministry of Agriculture and Agrarian Reform, Aden, Yemen. (In Arabic)

Ba-Angood, S.A.S. 1982a. *Research conclusions and recommendations for major crop pests in PDR Yemen for the period 1970-82.* Faculty of Agriculture, University of Aden, Aden, Yemen.

Ba-Angood, S.A.S. 1982b. Control of cotton bollworms in PDRY. *In* K.L. Hoeng, ed. *Proceedings of the International Conference on Plant Protection in the Tropics*, p. 581-588. Kuala Lumpur, Malaysia, Malaysian Plant Protection Society.

Ba-Angood, S.A.S. 1990. Preliminary survey of some natural enemies of major insect pests in PDRY. *Yemen: Studies & Research*, 2(2): 22-37.

Ba-Angood, S.A.S. 1991. *Cotton pests, diseases and weeds and the recommended methods of control in the united Yemen.* Faculty of Agriculture, University of Aden, Aden, Yemen. (In Arabic)

Ba-Angood, S.A.S. 1994. *Termite problems in Tihama with particular reference to oilseed crops and recommendations for control strategies.* AREA/AOAD/ UNDP. Oil Seed Crops Development Programme. RAB/89/024/A/01/99.

Ba-Angood, S.A.S., Abdessattar, M.A. & Masoud, H. 1985. *Plant protection in PDRY.* A country report presented to the consultation meeting on integrated pest management and rationalization of use of chemical pesticides, 16 to 20 September 1984, Algiers, Algeria. p 181-207. (In Arabic)

Farbrother, H.G. 1962. Crop water use studies at El-Kod. London, UK, Cotton Research Corporation. (Unpublished report)

Kamal, M. & Aghbari, A.A. 1985. *Manual of plant diseases in the Yemen Arab Republic.* Agricultural Research Authority, Protection Press.

Kassasian, K. 1980. Weed consultancy to PDR Yemen, 8 February to 2 April 1980. Rome, Italy, FAO.

MAWR. 1991. *Agricultural Statistics for 1990.* Statistics Project, Ministry of Agriculture and Water Resources, Yemen.

MAWR. 1993. *Agricultural Statistics Year Book 1992.* Ministry of Agriculture and Water Resources, Yemen.

Muallem, Ab.S. 1968. *Historical background information of the Agronomy Section.* El-Kod Agricultural Research Centre, Department of Agriculture, Ministry of Agriculture and Agrarian Reform, Aden, Yemen.

Muallem, Ab.S. 1982. *Conclusions and recommendations of research activities on major field crops in PDRY during the period 1970-1982.* Ministry of Agriculture and Agrarian Reform, Aden, Yemen.

Muallem, Ab.S. 1985. *Practical methods for cottonseed production in Democratic Yemen.* Training courses for technical assistants, Giar Training Centre, Yemen. (In Arabic)

Nasseh, O.M. & Mahyoub, A. 1987. *Revised list of insects found in YAR.* Yemen/German Plant Protection Project, Sana'a, Yemen.

Ogborn, J.E.A. 1960. *Progress reports from experiment stations 1960-61. South Arabia.* London, UK, Empire Cotton Growing Corporation.

Rijks, D.A. 1965. The use of water by cotton crops in Abyan, South Arabia. *J. Appl. Ecol.,* 2: 317-343.

Zaazou, M.H., Mahfod, S., Ba-Angood, S.A.S. & Al-Sagaff, A. 1976. *The control of cotton bollworms in PDR Yemen.* Technical Report No. 2. UNDP/FAO Agricultural Research and Training Project, El-Kod & Giar, PDY/71/516.

Part II

Invited papers

Important cotton diseases in the Near East: challenges and solutions

Kamal M. El-Zik

INTRODUCTION

Many diverse economic and social pressures are stimulating a re-evaluation of today's agriculture. Growers across the United States cotton belt and in many other cotton-producing countries, are facing intense economic pressures as prices for their products hold steady while costs of production increase. At the same time growers and the general public are becoming increasingly concerned about the environmental effects associated with cotton production, particularly fertilizers and pesticides as they relate to food safety, pest resistance to pesticides, surface and groundwater contamination and environmental pollution.

The impressive increases in cotton productivity achieved during this century have resulted from genetic improvements in crop cultivars and from advances in agricultural technology and management practices. However, cotton growers around the world still face many problems in producing the crop. Most of these problems are related to cotton pests, their management and control. These problems are challenges and solutions to them must be found through research and development.

One of the most significant constraints to sustaining or increasing cotton productivity and quality is loss to pest attack, including pathogens that cause disease. In the United States, estimates of annual yield losses caused by all diseases of cotton, over the 33-year period 1952 to 1984, ranged from 8 to 18 percent, with an average of 12.7 percent (El-Zik, 1985). In African and Asian countries yield losses of 10 to 30 percent are often experienced as a result of bacterial blight (El-Zik and Thaxton, 1994; Verma, 1986). Yield losses may be as high as 50 to 70 percent in severe epidemics, where extensive leaf, stem and boll infection occur. Damage caused by leaf curl virus and whitefly can be devastating.

Substantial progress in controlling cotton diseases has been achieved because research has provided a basic understanding of diseases and their etiology, epidemiology, pathogen variability, host resistance and resistance mechanisms, host-pathogen-environment interactions and management strategies. More research is needed into the control of cotton diseases. There are important factors in the breeding and culture of the cotton plant, as conditioned by its environment, that are common to the suppression and control of insects, plant pathogens, nematodes and weeds. These common tactics serve to link the suppression of different pest classes into an integrated management strategy. A multidisciplinary team approach is required to develop a basic and applied research base. Integrated pest management (IPM) systems are rapidly evolving into integrated crop management systems (ICMS) because of the increasing difficulty in separating the control of pests from the culture of the crop (El-Zik and Frisbie, 1985; El-Zik, Grimes and Thaxton, 1989). Successful IPM and ICMS programmes utilize and integrate resistant cultivars, cultural management tactics, biological agents and selective use of chemicals.

DISEASE MANAGEMENT STRATEGIES
Host plant resistance
Genetic resistance is one of the oldest methods of pest control and the most effective defence against cotton pests, although it is recognized that the usefulness of a new cultivar will depend primarily on its yielding ability and fibre and seed quality. Resistant cultivars provide the cornerstone for a successful IPM system. The cultivar sets the framework for the level of susceptibility to pests, the tactics applied to manage the crop, production costs and net profit (El-Zik, 1985). Resistant cultivars, even those with only moderate levels of resistance to pests, are highly compatible with all other control tactics; they contribute to stability and offer advantages to IPM, ICMS and sustainable agriculture. Resistance may either be a contributing factor or the primary means for controlling pests. Genetic resistance is usually used together with other pest control methods, including cultural,

biological and chemical approaches. Resistant cultivars may not require as many treatments or such high rates of pesticide application to achieve adequate pest control as susceptible cultivars.

A key consideration in modern agriculture is maintaining the health of plants throughout the growing season, so that they may approach the full genetic potential of both yield and quality of fibre and seed (El-Zik and Thaxton, 1989). Crop health describes the relative freedom of plants from biotic and abiotic stresses. Improvement of cotton for resistance to only one or a few pests is not sufficient for sustainable agriculture. The most effective approach is breeding for multi-adversity resistance, that is, resistance to insects, pathogens and abiotic stresses (El-Zik and Thaxton, 1989).

Cultural practices

Over the centuries many cultural and management practices have been developed for the protection of cotton crops. Such practices include selection of planting sites; rotation of crops to discourage the buildup of damaging pest populations (common rotations are with small grains, maize or legumes); and tillage to destroy weeds, overwintering insects and pathogen inoculum. A balanced nutrition of the major elements (nitrogen, phosphorus and potassium) and minor elements is important in minimizing plant stress and susceptibility to pests and in maintaining plant health. The amount and form of fertilizer, especially of nitrogen fertilizers, and time of application are critical factors in reducing disease epidemics (El-Zik and Frisbie, 1985). Excess nitrogen generally enhances pest problems. Other cultural practices include: cultivar selection; planting dates to avoid periods that are liable to high levels of damage; plant population; water management – both the timing and the amount; the use of plant growth regulators and harvest aid chemicals; time of harvest; crop residue management; trap crops; soil solarization; and regulatory programmes (El-Zik, 1990; El-Zik and Frisbie, 1985; El-Zik, Grimes and Thaxton, 1989; Watkins, 1981). Some of these cultural and management practices provide for the biological destruction and/or suppression of disease organisms.

Biological control

Biological control aimed directly at the pathogen or mediated through adjustments in the host offers unlimited opportunities to reduce losses caused by diseases. Antibiosis is the most widely recognized mechanism that may cause inactivation or destruction of soil-borne plant pathogen propagules (Cook and Baker, 1983). A number of actinomycetes, bacteria and fungi have been isolated from soil and plant surfaces and some have shown potential as biological control agents. Recent research has focused on utilizing microbes isolated from the cotton rhizosphere and rhizoplane as biological agents to control seed and seedling pathogens (Sterling, El-Zik and Wilson, 1989). Some species of *Gliocladium*, *Trichoderma*, *Pseudomonas* and *Bacillus* show promise for seed and seedling disease control (El-Zik, 1990; Sterling, El-Zik and Wilson, 1989).

Chemical control

Fungicides are used effectively to control seed and seedling pathogens. The use of fungicides has been the only practical way to control southwestern cotton rust. Several chemical fumigants (methyl bromide, miban, chloropicrin and dichloropropene) are effective in eradicating verticillium and fusarium wilts in soils. Effective treatment, however, frequently requires large doses of the chemical and deep placement, which is very expensive. Chemicals may also have adverse effects on beneficial organisms, which may lead to decreased yields. Several nematicides are used to control nematodes. There is a need for new and safer fungicides that can be used at reduced rates and targeted to specific pathogens.

IMPORTANT COTTON DISEASES

Some of the most important cotton diseases are widely distributed throughout the world and cause loss in yield, fibre quality and profit. The seed and seedling pathogens, bacterial blight, verticillium and fusarium wilts and leaf curl virus are common and important diseases that occur in most of the Near East cotton-producing regions. The twentieth-century agricultural revolution and advancements in technology have disrupted and complicated the stability of pests and beneficial species in the agro-ecosystem.

An integrated pest and crop management system is essential to minimize losses from diseases, since no single method is highly effective in controlling these pathogens.

Seed and seedling diseases

Cottonseed that has the inherent traits of good germination, resistance to seed and seedling pathogens and production of healthy seedlings when planted early in the season under cool, moist conditions is essential for efficient cotton production. A uniform and healthy stand with vigorous and evenly distributed seedlings is the foundation of cotton production. Seed and seedling diseases are the main cause of poor final stands and usually result in yield loss. Substantial improvement in resistance to seed and seedling pathogens and in the development of traits contributing to stand establishment and vigour have been achieved in modern cotton cultivars (El-Zik and Thaxton, 1989; 1992).

Pathogens and disease syndrome. The main pathogens causing seed and seedling diseases are *Fusarium* and *Pythium* spp., *Rhizoctonia solani* Kuhn, and *Thielaviopsis basicola* (Berk.) & Br. (El-Zik and Thaxton, 1992; Minton and Garber, 1983; Watkins, 1981). Disease syndrome includes seed rot, pre- and post-emergence damping-off and seedling root rot.

Factors affecting the disease complex. Stand establishment and vigour in cotton are the end results of a complex phenomenon that involves biotic, abiotic, environmental and management factors and their interactions (El-Zik and Thaxton, 1992). Figure 5 represents a conceptual model of these factors. Stand and vigour are affected by host resistance, seed viability and germination, emergence and growth, pathogen virulence and inoculum density. Other factors affecting the host and the pathogens are temperature, soil moisture and planting depth.

Stand and vigour are also influenced by seed germination rate, emergence, tissues affected by primary and secondary pathogens, which cause seed rot and pre- and post-emergence damping-off (Figure 6). Chemical injury caused by herbicides or fertilizers reduces stand and predisposes seedlings

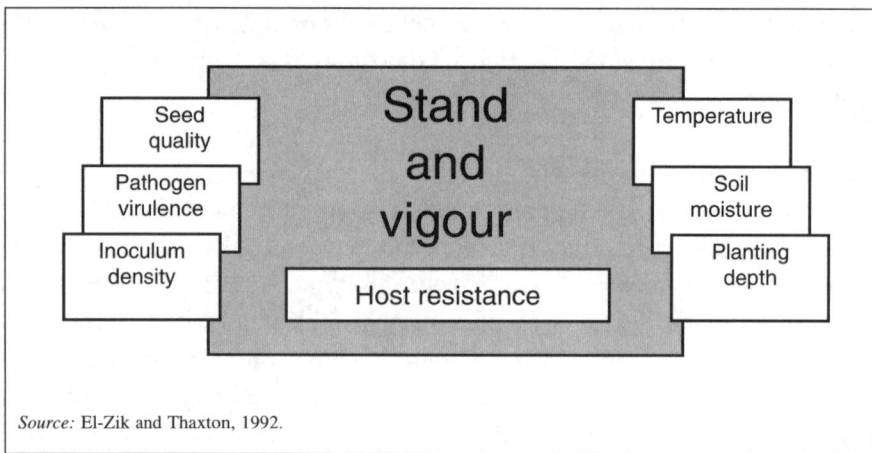

Source: El-Zik and Thaxton, 1992.

FIGURE 5
Conceptual model for host, pathogen and environmental components affecting cotton stand and vigour

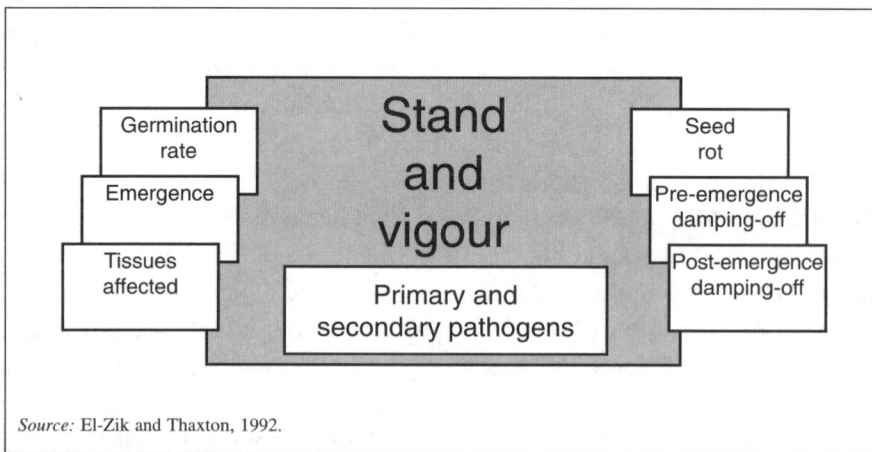

Source: El-Zik and Thaxton, 1992.

FIGURE 6
Seed and seedling factors affecting stand and vigour

to infection. These factors individually and collectively affect stand and vigour.

Resistance to the disease complex. Resistance to cotton seed and seedling pathogens requires genetic changes in several traits including: seed and seedling performance in cool soil, preservation of seed quality, seed and seed coat resistance to pathogens and resistance of tissues affected (root, epicotyl and hypocotyl) to specific pathogens (El-Zik and Thaxton, 1989; 1992).

One of the most important factors affecting stand establishment and vigour is the quality of the planting seed. Reduced quality is the result of deterioration of the seed while it is in the field or in storage. Deterioration of cottonseed in the field usually occurs between boll opening and harvesting when seeds undergo weathering through exposure to moisture and heat. Susceptibility to the seed and seedling pathogens increases with increased deterioration of seed.

Integrated control. Control strategies for seed and seedling diseases include cultivar selection, seed quality and planting date. Factors affecting stand ability at sowing are temperature, moisture, seed-bed preparation, seed depth and seeding rate. Fungicides are used effectively to control seed and seedling pathogens. Coating seed with two or more fungicides (including systemics) prevents the carryover of pathogens on the seed and reduces both pre- and post-emergence diseases. Applying soil fungicides as planter-box or in-furrow treatments during sowing is a most effective means of controlling seedling disease in areas where the disease is severe (Minton and Garber, 1983).

The biocontrol agents *Bacillus subtilis* (Hagedorn, Gould and Bardinelli, 1985), *Trichoderma* spp. (Elad, Kalfon and Chet, 1982), *Gliocladium virens* (Howell, 1982) and *Pseudomonas fluorescens* (Howell and Stipanovic, 1979; 1980) have been field tested extensively and show promise for controlling *Rhizoctonia solani* and *Pythium ultimum*. *B. subtilis* GBO3 is being used commercially with fungicides to control cotton seed and seedling pathogens. Biological control offers a means to improve the health and enhance the productivity of plants by suppression or destruction of pathogen inoculum, protection of plants against infection or increasing the ability of plants to resist pathogens.

Bacterial blight

Bacterial blight of cotton is caused by *Xanthomonas campestris* pv. *malvacearum* (E.F. Smith) Dye (*Xcm*) and occurs in most cotton-producing regions of the world. The disease can affect all plant parts in the form of seedling blight on seedlings, angular leaf spots on leaves, blackarm lesions on stems and petioles and boll rot (Bird, 1986; El-Zik and Thaxton, 1994; Hillocks, 1992; Verma, 1986). The disease is potentially very destructive where wind-driven rain or sprinkler-irrigation disperses the pathogen. Bacterial blight of cotton is one of the most studied diseases and comprehensive reviews have been provided by researchers (Brinkerhoff, 1963; 1970; El-Zik and Thaxton, 1994; Hillocks, 1992; Innes, 1983; Knight, 1957; Lagiere, 1959; Verma, 1986; Wickens, 1953).

Epidemiology. The main sources of primary inoculum are contaminated crop residues and infected seed. The pathogen can survive in and on the seed. Rain splash, wind, insects and field equipment spread the bacteria and cause secondary infection. Environmental conditions that are conducive to disease development include periods of heavy, wind-driven rain after the canopy has formed, relative humidity greater than 85 percent within the canopy, and high temperatures of 34 to 38°C during the day and 17 to 20°C at night during the secondary phase of the disease.

Pathogen variability. The genetic variability of the pathogen for virulence and other traits has been well documented (Brinkerhoff, 1963; 1970; Cross, 1963; Follin, 1981; 1983; Verma, 1986). Soon after the cultivation of blight-resistant cottons began in the United States, new and virulent races of the pathogen appeared. The Cotton Disease Council currently recognizes 19 United States races of *Xcm* based on disease reactions on a set of ten upland *Gossypium hirsutum* host differential strains. In Burkina Faso several strains of *Xcm* appeared that were virulent on the entire set of host differentials

(Follin, 1981; 1983). A survey of races present in western and central Africa revealed that a population exists that can overcome all the major genes for bacterial blight resistance currently used in breeding programmes in the United States. These new isolates were designated HV1, HV3, HV7, Chad and Sudan.

A recent study on the variability and virulence of *Xcm* races showed that races 1 and 2 and isolate HV3 were the least virulent (Wallace and El-Zik, 1992). Races 7 and 10 and isolates HV7 and Sudan were intermediate in virulence and *Xcm* race 18 and isolate HV1 were the most virulent.

World distribution of **Xcm** *races.* It is recognized that symptom expression in any one interaction between pathogen and host is greatly influenced by environmental factors, host nutrition and plant age. The United States set of host differentials has been used in several countries to determine the races present in the local population of the pathogen (Table 42). In most countries, more than one race has been identified. Race 1 was reported from Australia, India and the United States. Races 2 to 5 were reported from the United States and India, race 6 from Nigeria, Zimbabwe and India. Race 18, the most virulent *Xcm* race in the United States, was also reported from Australia, west and central Africa, India, Pakistan and Nicaragua (El-Zik and Thaxton, 1994).

Genetics of resistance. Heritable resistance to bacterial blight was first demonstrated in the Sudan in 1939 (Knight and Cloustan, 1939). At least 22 major genes for resistance have been reported. The majority of genes identified have been described as partially to completely dominant for resistance. The effectiveness of specific "*B*" genes and gene combinations in conferring resistance and the effect of the genetic background and modifier genes on enhancing resistance have been reported (Bird, 1986; El-Zik and Bird, 1970; Follin *et al.,* 1988; Innes, 1974; 1983; Knight, 1957; Wallace and El-Zik, 1989). Quantitative analysis of resistance to bacterial blight indicated additive, dominance and epistatic gene action. Single-resistance genes confer relatively high levels of resistance to a few races (vertical resistance), but are vulnerable to the other races of the pathogen.

TABLE 42

Races of bacterial blight pathogen and countries where they have been identified

Race no.	Countries and regions where identified
1	Australia, United States, India
2	Australia, United States, India
3	United States, India
4	United States, India
5	Australia, United States, India
6	Nigeria, Zimbabwe, India
7	Australia, Venezuela, Nigeria, United States
8	Venezuela, Peru, Nigeria, Pakistan, India
9	Australia
10	Pakistan, Nigeria, Zimbabwe, India, United States
11	United States, India, Nigeria
12	Pakistan, United States, India
13	India
14	Philippines, United States, India
15	United States, India
16	West and Central Africa, Nigeria, India
17	India
18	Australia, West and Central Africa, Pakistan, Nicaragua, India, United States
19	Brazil, India

Thus, different combinations of single "*B*" genes and modifiers are necessary to obtain a stable source of resistance (horizontal resistance).

Selection must be made utilizing a compatible race mixture of the pathogen, including virulent races, in order to identify gene combinations that give broad spectrum resistance to many races (Bird, 1982; Brinkerhoff *et al.*, 1984; El-Zik and Thaxton, 1989; 1992; Thaxton and El-Zik, 1993; Verma, 1986). The race or mixture of races used in screening for resistance to *Xcm* in different countries depends on the races present in that location and the germplasm being worked with (El-Zik and Thaxton, 1994). Recurrent selection and the back-cross method have been used to develop horizontal resistance to all races of the pathogen (Brinkerhoff *et al.*, 1984; El-Zik and Thaxton, 1989; 1994).

Integrated control. An integrated approach is essential to control bacterial blight in areas where conditions are favourable to the development of epidemics. Cultural practices include chopping crop residues and burying them in moist soil, crop rotation or irrigation/flooding after crop residues have been ploughed in, sanitary practices during ginning and processing, the use of acid-delinted and fungicide-treated seed obtained from blight-free areas and furrow irrigation instead of sprinklers.

Fungicides, including seed treatment and foliar sprays, are used in Africa to protect the crop from epidemics. Bronopol, TCMTB, carboxin, copper compounds and oxycarboxin are used as seed treatments. Foliar sprays include the use of antibiotics (streptomycin sesquisulphate), carboxin or oxycarboxin.

Cotton pathologists and breeders have made considerable advances in breeding for resistance to bacterial blight. High resistance is available in several cultivars throughout the world, including the Siokras in Australia; Barakat, Bar, and Baras in the Sudan; Allen, UK74, UK71, Albar and Reba B50 in East and Central Africa; BJR-734 in India; and the Tamcot cultivars and MAR germplasm in the United States.

Verticillium wilt

Verticillium wilt, a vascular wilt disease affecting cotton, is caused by *Verticillium dahliae* (Kleb.). The disease is recognized as a serious threat to cotton production in several countries around the world where cotton is grown under relatively cool conditions. Verticillium wilt has become a major disease problem of irrigated cotton in Turkey, the Syrian Arab Republic, the Islamic Republic of Iran and northern Iraq. Losses from the disease range from 2 percent to as high as 30 percent annually (Bell, 1992).

Epidemiology. The fungus survives as dormant microsclerotia in soil and decomposed plant debris for several years and is stimulated to grow by root exudates. There is a time interval of about 14 days between root infection and detection of disease symptoms in leaves. Two major strains of *V. dahliae* differ in pathogenesis – one causes the common leaf symptoms and the other can cause complete defoliation and the shedding of small bolls. Temperature is the most important environmental factor. Optimum temperatures for disease severity are 22 to 27°C (Bell, 1992; El-Zik, 1985). Under optimal conditions for infection, the susceptible host is normally killed by a combination of toxic fungal metabolites, accumulated fungal material and host response to infection, which leads to vascular occlusion and moisture deficit.

Integrated management system. A conceptual model of an integrated management system for controlling verticillium wilt has been developed (Figure 7). The model integrates and links our knowledge of system components (host resistance, genetic potential, fruit set, other pests in the system, pathotypes and inoculum density of the pathogen, disease progress, temperature and moisture and rhizoplane and phylloplane environment) to plant health, cotton yield and quality (El-Zik, 1985). The system adjusts biological activities according to environment and management tactics. Management practices and options to reduce disease severity and loss include increasing crop rotation, sanitation, plant density and soil temperature and decreasing crop residues, weeds, nitrogen rate and irrigation frequency and amount. No single method is highly effective in controlling verticillium wilt so an integrated management system is necessary to minimize losses from the disease.

Control measures. The most effective control is achieved by growing adapted resistant cultivars and using cultural and management practices known to reduce disease severity (Bell, 1992; El-Zik, 1985). Control methods to prevent a disease epidemic should be aimed at reducing the initial inoculum density in the soil, the survival and dispersal of inoculum, the rate of infection and the time period over which the crop is exposed to infection. Control measures may be implemented before, during or after sowing, depending on the amount of inoculum in the soil (Figure 8).

If the initial inoculum density is high, control measures may not lower the number of infective propagules enough to reduce disease severity and maintain yield. If inoculum density is 22 propagules per gram of soil, or higher, the main option is crop rotation. If inoculum density is five to 20 propagules per gram, several practices presowing, during sowing and post-sowing are available (El-Zik, 1985). Presowing practices include sanitation, site selection, weed control, reduced row spacing and fertilization. Soil

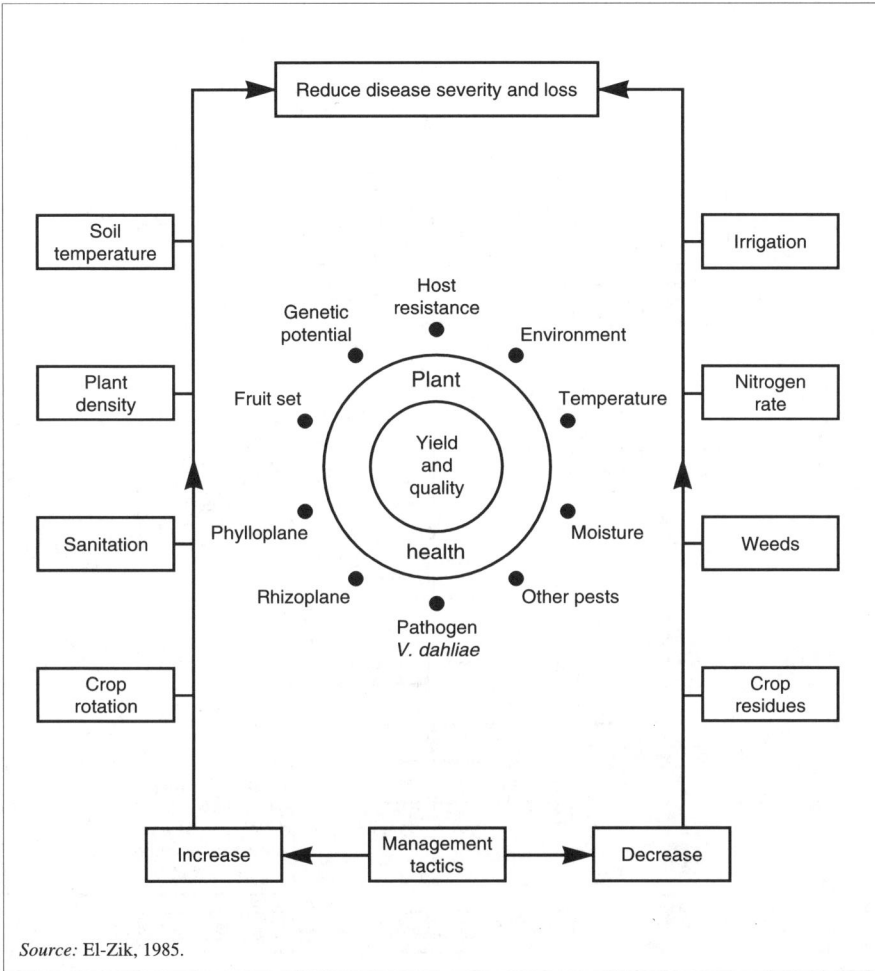

FIGURE 7
Conceptual model of an integrated management system for controlling verticillium wilt of cotton

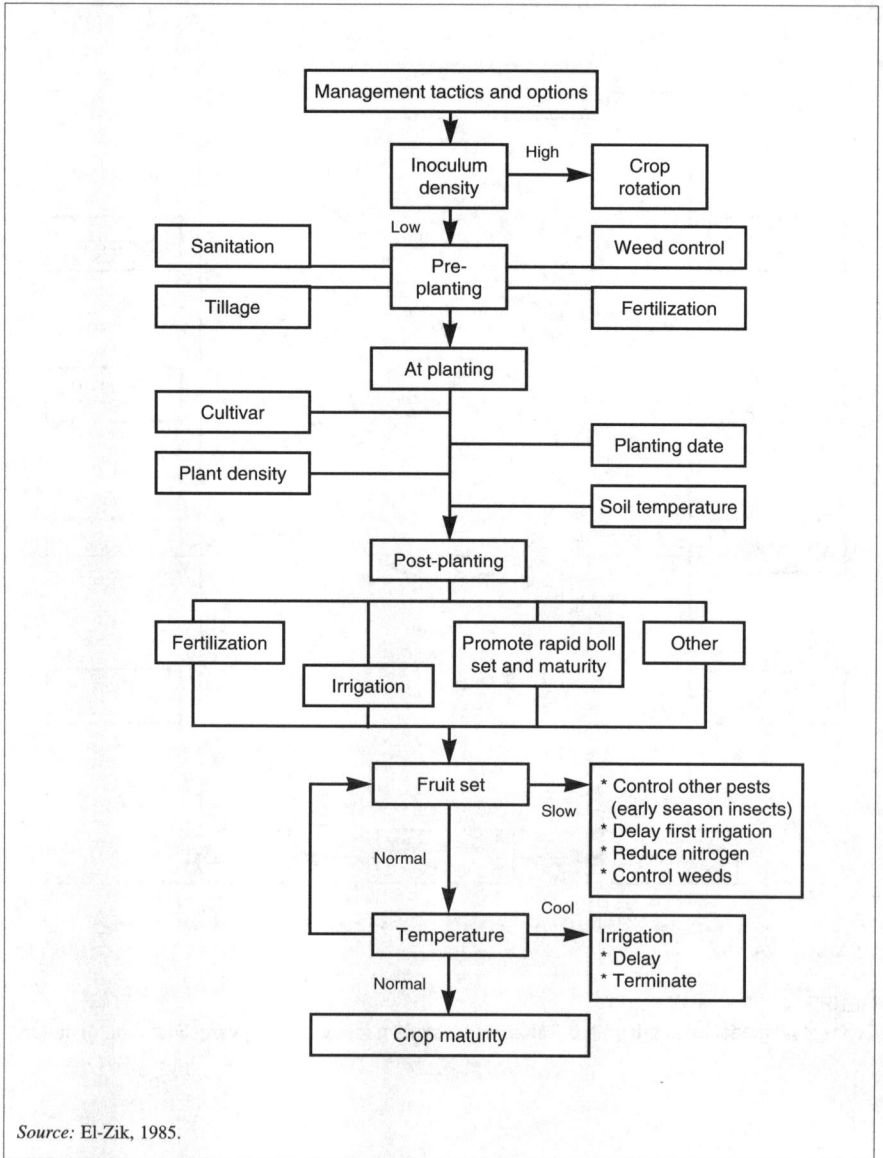

Source: El-Zik, 1985.

FIGURE 8
Integrated seasonal management tactics and options to control verticillium wilt of cotton

solarization increases the soil temperature under clear polyethelene sheeting and has been shown to decrease the population of *V. dahliae*.

Practices during sowing include management tactics that raise soil temperature, the use of optimum sowing dates and increased plant density and the choice of a resistant cultivar. Examples of resistant cultivars are the Allepos in the Syrian Arab Republic, Nazilli 66-100 and Nazilli 84 in Turkey and the Acalas in the United States.

Post-sowing practices include fertilizer rate and source, irrigation timing and amount and the promotion of rapid boll set and early maturity. The stage of cotton plant development when foliar symptoms of verticillium wilt occur has a direct effect on cotton lint yield and quality. If the disease occurs early, about 60 days after sowing, it causes reduction in plant growth and development, square shed and possibly small boll shed, which are the most significant factors limiting yield (El-Zik, 1985). At the onset of flowering, about 80 days after sowing, rapid pathogenesis occurs, which is directly related to the diversion of a portion of photosynthate to reproductive structures. Chemical control can be effective, but is not generally used because of its prohibitive cost.

Fusarium wilt

Fusarium wilt, caused by *Fusarium oxysporum* Schlecht f. sp. *vasinfectum* Atk. Sny. & Hans occurs in several Near Eastern countries. It is a soil-borne pathogen and is also carried within the seed (Hillocks, 1992). The pathogen causes damage to cotton and, together with the root knot nematode, *Meloidogine incognita*, can destroy it. The pathogen is favoured by temperatures of above 23°C (optimum 30 to 32°C) and light soils of neutral to acid pH. Symptoms usually begin four weeks after sowing and plants become more susceptible at flowering. Six races of the pathogen are recognized; races 1 and 2 originated in the United States, race 3 is found in Egypt, race 4 in India, race 5 was identified in the Sudan and race 6 in Brazil (Hillocks, 1992).

Integrated control. Progressive improvements have been made in several countries in the development of cultivars with increasing levels of resistance to both fusarium wilt and root knot nematode. Rotations with barley, wheat, maize, mustard and clover reduce the incidence of wilt the following season. High rates of ammonium nitrogen tend to increase fusarium wilt incidence. Soil solarization has been shown to decrease disease incidence, but the fungus is less affected by solarization than *Verticullium dahliae* is. Certain soils are known to be suppressive to fusarium wilt because of the presence of antagonistic microorganisms such as *Azotobacter*, *Bacterium agile* and *Trichoderma harzianum* Rifai (Cook and Baker, 1983; Hillocks, 1992).

Populations of the fungus can be decreased by using, for example, a mixture of chloropicrin and methyl-bromide fungicides, but the cost may not give an economic return. In fields where fusarium is associated with root knot nematode, disease incidence can be decreased by controlling the nematodes with nematicides.

Cotton leaf curl

The cotton leaf curl gemini-virus (Mansoor *et al.*, 1993; Nelson, Orum and Nadeem, 1994) is transmitted by the whitefly *Bemisia tabaci* L. (Genn.) (Hemiptera: Aleyrodidae), which is the only known vector. The serious spread of the disease in Pakistan, the Sudan and other countries has been caused by the use of highly susceptible cultivars of cotton, significant changes in whitefly population dynamics and seasonal distribution and the use of non-selective insecticides. The greatest damage and subsequent yield loss occurs when cotton is infected by the leaf curl virus at early growth stages; late-season infections result in only minimal damage. Under conditions of early infection up to 70 percent yield losses have been reported. Climatic conditions, including rainfall, wind and temperature, affect the epidemiology of leaf curl virus. Regional spatial and temporal distributions of infection hazards for the virus disease can be characterized by the use of geographic information systems (GIS) and geostatistics (Nelson, Orum and Nadeem, 1994).

Integrated control. Growing tolerant or resistant cultivars, eliminating cotton stubble and rigorously controlling insect vector populations and alternate hosts that serve as known or suspected virus reservoirs, are critical for controlling leaf curl virus. Minimizing serious outbreaks of cotton leaf curl virus requires the integrated control of whitefly. There are a number of ways by which whitefly problems can be reduced, including the destruction of crop residues which harbour whitefly populations, not sowing cotton next to other crops already infested with whitefly, the separation of plots spatially or temporally, early sowing to minimize the effect of subsequent population buildup and using spraying equipment that gives good underleaf cover. Insecticides should be used selectively, on the basis of economic thresholds, to preserve beneficial insects, and should be rotated to avoid the development of whitefly resistance to insecticides.

THE FUTURE
Developments in molecular biology, biotechnology and systems analysis provide new opportunities for furthering efforts to develop cultivars resistant to pests and for managing and controlling pests. Recent technical advances in recombinant DNA and tissue culture methods have developed to a point where biotechnology will have an impact on the efficiency and success of conventional plant breeding. Biotechnology will assist breeders to improve crops by increasing the efficiency and effectiveness of selection, broadening the genetic base from which the breeder can select new cotton genes (through direct or indirect genetic modification) and introducing novel genes to control pests. New genes for the expression of the *Bacillus thuringiensis* toxin and for herbicide resistance have already been introduced.

The main components of crop biotechnology are DNA markers, agriculturally important genes, gene transfer and regeneration and stable and safe transgenic plants. Recent advances in biotechnology are seen to offer the possibility of revolutionary new solutions in the area of crop protection. Genetic modification and engineering have great potential to contribute to more sustainable and environmentally sound agricultural systems.

Systems analysis, including conceptual and simulation models, and decision support systems or expert systems, are important for crop and pest management, because they establish thresholds for pests and integrate system components. Genetic algorithms, which are computer-based techniques for exploring search domains in a manner analagous to natural selection, can be useful in breeding programmes.

As the twenty-first century approaches, many new challenges and opportunities arise. The challenges are to mould and adapt new technologies to create and engineer future cotton cultivars and to develop integrated pest and crop management systems that will provide healthy, efficient and safe plants to supply the fibre and food for an ever-increasing world population.

References

Bell, A.A. 1992. Verticillium wilt. *In* R.J. Hillocks, ed. *Cotton diseases,* p. 67-126. Wallingford, UK, CAB International.

Bird, L.S. 1982. The MAR (multi-adversity resistance) system for genetic improvement of cotton. *Plant Dis.,* 66: 172-176.

Bird, L.S. 1986. Half a century dynamics and control of cotton diseases: bacterial blight. Cotton Disease Council, *Proc. Beltwide Cotton Res. Conf.,* 46: 41-48.

Brinkerhoff, L.A. 1963. Variability of *Xanthomonas malvacearum* – the cotton bacterial blight pathogen. *Okla. Agric. Exp. Sta. Tech. Bull.,* T-98. 96 pp.

Brinkerhoff, L.A. 1970. Variation in *Xanthomonas malvacearum* and its relation to control. *Annu. Rev. Phytopathol.,* 8: 85-110.

Brinkerhoff, L.A., Verhalen, L.M., Johnson, W.M., Essenberg, M. & Richardson, P.E. 1984. Development of immunity to bacterial blight of cotton and its implications for other diseases. *Plant Dis.,* 58: 168-173.

Cook, R.J. & Baker, K.F. 1983. *The nature and practice of biological control of plant pathogens.* St. Paul, MN, USA, The American Phytopathological Society. 539 pp.

Cross, J.E. 1963. Pathogenicity differences in Tanganyika populations of *Xanthomonas malvacearum. Emp. Cotton Grow. Rev.,* 40: 125-130.

Elad, Y., Kalfon, A. & Chet, I. 1982. Control of *Rhizoctonia solani* in cotton by seed coating with *Trichoderma* spp. spores. *Plant Soil,* 66: 279-281.

El-Zik, K.M. 1985. Integrated control of verticillium wilt of cotton. *Plant Dis.,* 69: 1025-1032.

El-Zik, K.M. 1986. Half a century dynamics and control of cotton diseases:

dynamics of cotton diseases and their control. Cotton Disease Council, *Proc. Beltwide Cotton Prod. Res. Conf.,* 46: 29-33.

El-Zik, K.M. 1990. Concepts and achievements of IPM in cotton disease management. Cotton Disease Council, *Proc. Beltwide Cotton Prod. Res. Conf.,* 50: 15-19.

El-Zik, K.M. & Bird, L.S. 1970. Effectiveness of specific genes and gene combinations in conferring resistance to races of *Xanthomonas malvacearum* in upland cotton. *Phytopathol.,* 60: 441-447.

El-Zik, K.M. & Frisbie, R.E. 1985. Integrated crop management systems for pest control and plant protection. *In* N.B. Mandava, ed. *Handbook of natural pesticides: methods. Vol. I: theory, practice and detection,* p. 21-122. Boca Raton, Fl., USA, CRC Press.

El-Zik, K.M., Grimes, D.W. & Thaxton, P.M. 1989. Cultural management and pest suppression. *In* R.E. Frisbie, K.M. El-Zik & L.T. Wilson, eds. *Integrated pest management systems and cotton production,* p. 11-36. New York, USA, John Wiley & Sons.

El-Zik, K.M. & Thaxton, P.M. 1989. Genetic improvement for resistance to pests and stresses in cotton. *In* R.E. Frisbie, K.M. El-Zik & L.T. Wilson, eds. *Integrated pest management systems and cotton production,* p. 191-224. New York, USA, John Wiley and Sons.

El-Zik, K.M. & Thaxton, P.M. 1992. Breeding for resistance to seed-seedling and bacterial blight diseases of cotton. Cotton Improv. Conf. *Proc. Beltwide Cotton Prod. Res. Conf.,* 44: 560-563.

El-Zik, K.M. & Thaxton, P.M. 1994. Breeding for resistance to bacterial blight of cotton in relation to races of the pathogen. *In* G. Constable & N. Forrester, eds. *Proc. First World Cotton Research Conference,* 13 to 17 February 1994, Brisbane, Australia.

Follin, J.C. 1981. Evidence of a race of *Xanthomonas malvacearum* (E.F. Smith) Dow. which is virulent against the gene combination B_2B_3 in *Gossypium hirsutum* L. *Coton et Fibres Trop.,* 36: 34-35.

Follin, J.C. 1983. Races of *Xanthomonas campestris* pv *malvacearum* (Smith) Dye in Western and Central Africa. *Coton et Fibres Trop.,* 38: 277-280.

Follin, J.C., Girardot, B., Mangano, V. & Benitez, R. 1988. New results on inheritance of immunity to bacterial blight (*Xanthomonas campestris* pv *malvacearum* (Smith) Dye, races 18 and 20) in the cotton plant (*Gossypium hirsutum* L.). *Coton et Fibres Trop.*, 43: 167-174.

Hagedorn, D., Gould, W.D. & Bardinelli, T.R. 1985. Characterization of the bacterial populations associated with the cotton rhizoplane. Cotton Disease Council, *Proc. Beltwide Cotton Prod. Res. Conf.*, 45: 31-32.

Hillocks, R.J., ed. 1992. *Cotton diseases.* Wallingford, UK, CAB International.

Howell, C.R. 1982. Effect of *Gliocladium virens* on *Phythium ultimum*, *Rhizoctonia solani* and damping-off of cotton seedlings. *Phytopathol.*, 72: 496-498.

Howell, C.R. & Stipanovic, R.D. 1979. Control of *Rhizoctonia solani* on cotton seedlings with *Pseudomonas fluorescens* and with an antibiotic produced by the bacterium. *Phytopathol.*, 69: 480-482.

Howell, C.R. & Stipanovic, R.D. 1980. Suppression of *Phythium ultimum* induced damping-off of cotton seedlings by *Pseudomonas fluorescens* and its antibiotic, pyoluteorin. *Phytopathol.*, 70: 712-715.

Innes, N.L. 1974. Resistance to bacterial blight of cotton varieties homozygous for combination of B resistance genes. *Ann. Appl. Biol.*, 78: 89-98.

Innes, N.L. 1983. Bacterial blight of cotton. *Cambridge Philos. Soc. Biol. Rev.*, 58: 157-176.

Knight, R.L. 1957. Blackarm disease of cotton and its control. *Proc. Second Int. Plant Protection Conf.*, p. 53-59. London, UK Butterworth.

Knight, R.L. & Clouston, T.W. 1939. The genetics of blackarm resistance. I. Factors B_1 and B_2. *Genet.*, 38: 133-159.

Lagiere, R. 1959. *La bacteriose du cotonnier,* Xanthomonas malvacearum *(E.F. Smith) Dowson, dans la monde et en République Centrafricaine.* Paris, France, Institut de Recherches du Coton et des Textiles Exotiques. 252 pp.

Mansoor, S.I., Bedford, M. S., Pinner, J. Stanley & Markham, P.G. 1993. A whitefly transmitted geminivirus associated with cotton leaf curl disease in Pakistan. *Pak. J. Bot.*, 105-107.

Minton, E.B. & Garber, R.G. 1983. Controlling the seedling disease complex of cotton. *Plant Dis.*, 67: 115-118.

Nelson, M.R., Orum, T.V. & Nadeem, A. 1994. Regional analysis of epidemics of whitefly transmitted gemini viruses in cotton. Cotton Disease Council, *Proc. Beltwide Cotton Prod. Res. Conf.*, 54: 270-271.

Sterling, W.L., El-Zik, K.M. & Wilson, L.T. 1989. Biological control of pest populations. *In* R.E. Frisbie, K.M. El-Zik & L.T. Wilson, eds. *Integrated pest management systems and cotton production,* p. 155-189. New York, USA, John Wiley & Sons.

Thaxton, P.M. & El-Zik, K.M. 1993. Methods for screening and identifying resistance to the bacterial blight pathogen in cotton in the MAR program. Cotton Disease Council, *Proc. Beltwide Cotton Prod. Res. Conf.,* 53: 211-212.

Verma, J.P. 1986. *Bacterial blight of cotton.* Boca Raton, FL. USA, CRC Press Inc. 278 pp.

Wallace, T.P. & El-Zik, K.M. 1989. Inheritance of resistance in three cotton cultivars to the HVI isolate of bacterial blight. *Crop Sci.,* 29: 1114-1119.

Wallace, T.P. & El-Zik, K.M. 1992. Reaction of three cotton (*Gossypium hirsutum*) cultivars to single and mixed isolates of *Xanthomonas campestris* pv *malvacearum. Plant Pathol.,* 41: 569-572.

Watkins, G.M., ed. 1981. *Compendium of cotton diseases.* St. Paul, MN, USA, The American Phytopathological Society. 87 pp.

Wickens, G.M. 1953. Bacterial blight of cotton. A survey of present knowledge, with reference to possibilities of control of the disease in African rain-grown cotton. *Emp. Cotton Grow. Rev.,* 30: 81-103.

Integrated pest management
of cotton pests in Texas

P.L. Adkisson

INTRODUCTION

Cotton is the most important cash crop in Texas. In 1991, Texas planted 2.6 million ha of cotton and produced 4.8 million bales. Almost all of this was produced by American upland varieties (*Gossypium hirsutum*) except for several thousand bales of Pima (*G. barbadense*) produced under irrigation in West Texas. Texas now produces almost 50 percent of the cotton grown in the United States.

The control of insect pests has been a major problem for cotton producers in Texas since the first days of cotton production. The invasion of Texas by the boll weevil (*Anthonomus grandis* Boheman Coleoptera: Curculionidae) from Mexico in the late 1890s and its subsequent spread across the south and southeastern United States caused great economic damage and considerable geographic displacement of the crop. Almost 100 years later the boll weevil is still a key pest of cotton and its presence complicates the control of other arthropod pests of the crop. The necessity for controlling arthropod pests to protect yield, has led to large quantities of insecticide being used annually on cotton.

The abundance of insect pests and the heavy treatment of cotton with insecticides have resulted in major research resources throughout this century being directed to the development of more effective, less costly, methods of control. This has occurred not only in Texas but also across the United States cotton belt. The basic principles, practices and strategies for IPM have evolved from research on cotton. Cotton, more than any other crop, has been the arena in which IPM has developed. Agricultural scientists at Texas A&M University have been among the leaders in developing and implementing IPM on cotton and many other crops (Adkisson *et al.,* 1982).

THE BASIS FOR IPM ON COTTON IN TEXAS

The boll weevil, cotton fleahopper (*Pseudatomoscelis seriatus* [Reuter] Hemiptera: Miridae) and the pink bollworm (*Pectinophora gossypiella* [Saunders] Lepidoptera: Gelichiidae) are key pests of cotton in most of Texas. These pests must be controlled annually by cultural practices and insecticides to prevent large losses in yield. The bollworm (*Heliothis zea* Boddie Lepidoptera: Noctuidae) and the tobacco budworm (*Heliothis virescens* Fabricius Lepidoptera: Noctuidae) are major secondary pests. Populations of these two pests may attain large numbers when their natural enemies are killed by insecticides applied for control of the boll weevil and cotton fleahopper. (In Texas, the pink bollworm is kept below damaging numbers by cultural and phytosanitation practices and insecticides are seldom applied for its control.)

The basic strategy for IPM on cotton in Texas is obvious; control of the boll weevil and cotton fleahopper must be accomplished by tactics that conserve insect natural enemies and do not induce outbreaks of the bollworm and tobacco budworm.

In order to control the boll weevil and pink bollworm and cause the least damage to arthropod parasites and predators, a combination of measures are applied during the off-season, aimed at reducing the number of these pests to survive the winter. The measures include:

- early uniform sowing of early-maturing cotton varieties;
- management of fertilizer and irrigation (if used) to induce early maturation of the crop;
- use of desiccants and defoliants to terminate the crop and cause shedding of the fruit, which is suitable for boll weevil food and reproduction, and sustaining larvae of the pink bollworm.
- adding an effective organophosphate insecticide to the desiccant or defoliant to kill as many diapausing weevils as possible;
- harvesting of the crop as rapidly as possible;
- destroying stalks and ploughing under residues to kill diapause weevils and pink bollworm larvae; this includes destruction of the green and cracked bolls that may be cleaned out of mechanical harvesters at the end of rows.

These practices, when carried out by all cotton farmers over a large area, can reduce the numbers of weevils and pink bollworms to such low levels that they do not achieve crop damaging numbers in the following year. In Texas, planting and stalk destruction dates for cotton are established by law and strictly enforced (Adkisson *et al.*, 1982).

The cotton fleahopper is most damaging at the time the cotton begins to form flower buds (squares). Fortunately, there are several insecticides which used at low dosages will destroy enough fleahoppers to allow the cotton plants to fruit. The low dosages do not kill many of the insect's natural enemies and thus do not induce bollworm/budworm outbreaks. During this period, overwintering boll weevils may also be killed before they reproduce the first generation (Adkisson *et al.*, 1982; Frisbie, Hardee and Wilson, 1992).

Occasional pests include the cotton aphid, whitefly, thrips, red spider mite and *Lygus* spp. In several areas the cotton aphid and whitefly have become highly resistant to most insecticides and in some years may cause significant economic damage.

CROP MANAGEMENT AS A COMPONENT OF IPM

Cotton in Texas is rain-fed or given supplementary irrigation. Rain-fed cotton is usually produced in a short-season production system while irrigated cotton is generally produced in a longer, full-season system. Key features of the two systems, as characterized by Norman and Sparks (1991), follow.

Short-season cotton production

This production system relies chiefly on cultural techniques, including selection of short-season varieties, early sowing and optimal fertilization and irrigation. These practices shorten the production season and the period that cotton is vulnerable to insect attack. By permitting an earlier harvest, the system also greatly reduces the period of vulnerability to damage by adverse weather conditions. Short-season cotton varieties usually require 130 to 140 days from sowing to harvest if grown under optimal nitrogen and water conditions. These varieties fruit and mature more rapidly than traditional full-season varieties allowing earlier post-harvest stalk destruction and greater reduction of overwintering boll weevil and pink bollworm numbers.

The first 30 days of flowering are critical for an optimum, early boll set. The earliness factor in short-season production can be completely lost if damaging populations of insects occur as the first squares are formed. Heavy loss of early squares to overwintered weevils may also detract from short-season production. The boll weevil and the bollworm/tobacco budworm complex should be controlled with insecticides when they occur in damaging numbers. Because of the early maturity and quick fruiting of short-season cotton, field scouting should be intensified during this period to determine pest population levels and damage as well as beneficial insect abundance. Plant growth and fruiting rates should also be monitored to allow early detection of potential problems.

Full-season production
The full-season production system has been practised in Texas for many years. This system uses slower fruiting, indeterminate, longer maturing varieties grown with higher nitrogen inputs (greater than 33 kg per hectare) and abundant irrigation. The result is a long-season production period of 140 to 160 days from sowing to harvest. The system requires higher inputs and has proved to be a profitable method of cotton production in past years. Production costs have increased greatly in recent years, however. Increasing nitrogen fertilizer and amounts of irrigation water adds extra expense, prolongs the fruit development period and delays maturation. These factors expose the cotton to high populations of late-season pests such as the boll weevil, bollworm and tobacco budworm. This results in high production costs as multiple applications of insecticides are required to protect the crop throughout the longer fruiting period and, consequently, high yields must be obtained to offset the high production costs. The probability of crop loss from delayed harvest because of adverse autumn weather conditions is greater under this production system.

Full-season cotton varieties can be grown under a short-season production regime when soil types and rainfall allow. Early sowing in combination with reduced nitrogen (33 kg per hectare or less) and fewer irrigations, where applicable, result in a somewhat shorter production period.

TACTICS FOR SUPPRESSING INSECT PESTS DURING THE GROWING SEASON

The following are the recommendations made to Texas cotton growers by the Texas Agricultural Extension Service for the integrated control of cotton insects (Norman and Sparks, 1991; Parker, Huffman and Sansone, 1991).

Regular field monitoring is a vital part of any pest management programme because it is the only way reliable information can be obtained to determine if and when pest numbers reach the economic threshold. Monitoring involves more than just looking for insects. It should determine the insect density and damage level through the use of standardized, repeatable sampling methods. It is also a reliable way to gauge plant growth, fruiting, beneficial insect activity, weeds, diseases and the effects of pest suppression practices.

Control measures are needed when pest numbers reach a level at which further increases would result in excessive yield or quality losses. This level is known as the economic threshold or treatment level. The relationship between pest level, amount of damage and the ability of the cotton plant to compensate for damage, is greatly influenced by crop phenology and seasonal weather. The economic threshold is not constant but varies with factors such as the price of cotton, cost of control and stage of plant development.

Field inspections should be made every three to seven days when pests are present. When a cotton field is properly monitored, accurate and timely decisions can be made to optimize control efforts while minimizing risk.

Early-season pests

The early season is the first few weeks of the season from plant emergence up until the first third-grown square.

Field inspection and management of early-season insect pests are extremely important, particularly in a short-season production system. Loss of early squares may prolong the length of the growing season required to obtain adequate fruit set.

Cotton fleahopper. Adult fleahoppers move into cotton from host weeds when cotton begins to square. Both adults and nymphs suck sap from the tender portions of the plant, including small squares. Squares are susceptible to damage by fleahoppers from the pinhead size through to the third-grown stage.

Once cotton begins to produce the first small squares (the four- to six-leaf stage), the main stem terminal buds (about 8 to 10 cm of plant top) of 25 randomly selected plants at each of four or more locations across the field should be inspected for fleahoppers. During the first three weeks of squaring, if 15 to 25 cotton fleahoppers (nymphs and adults) are found in the 100 terminal samples, the economic threshold has been reached and economic damage may be expected. As plants increase in size and fruit load, larger populations of fleahoppers may be tolerated without yield reduction. Insecticides should not be applied just prior to or early in the flowering period as this will result in destruction of beneficial insects, possibly inducing an outbreak of bollworm and tobacco budworm.

Overwintered boll weevil. Overwintered boll weevils emerge from winter hibernation and enter cotton early in the growing season. They occur in very low numbers and females do not lay eggs until the first squares are about 0.6 cm in diameter (one-third-grown). Insecticides applied at this time will help suppress boll weevil population buildup until after peak flowering. This allows the plant to set a large number of bolls early, while having little adverse effect on numbers of mid- and late-season beneficial insects. Applications should not be made in fields where population buildup in past years has not occurred and where weevils are not found.

If weevils are found and the field has a history of heavy weevil infestation, early-season control applications may be economically feasible. The first application should be applied when squares are no more than one-third-grown. The second application should be applied three to five days later if weevils continue to move into the field. Treatment should be terminated at least two weeks before the initiation of flowering to avoid inducing bollworm/budworm outbreaks.

Mid- and late-season pests

The mid-season is the six-week fruiting period following the appearance of the first 0.6 cm diameter (one-third-grown) squares. Proper crop management and frequent field inspection for pests and beneficial arthropods will eliminate unnecessary insecticide applications during this period. These procedures ensure adequate fruit set and preservation of beneficial insects and spiders.

The late season is the remainder of the production season when the major objective is to protect immature bolls that can be expected to mature. Heavy irrigation and high rates of fertilizer prolong cotton plant growth and increase the chance of late-season insect damage. Short-season cotton production schemes minimize late-season insect problems.

Bollworm, tobacco budworm and boll weevil are the principal mid- and late-season insect pests. A major goal of a well-planned pest management programme (although not always achieved) is to avoid having to treat for bollworms and tobacco budworms. Indigenous parasites and predators and prolonged dry weather often suppress boll weevil, bollworm and budworm populations in the mid- and late seasons. For this reason, chemical control of the boll weevil during this period should be avoided if possible. If a satisfactory fruit set occurs during the first 30 days of flowering, higher numbers of weevil-damaged squares can be tolerated without loss of yield. However, if fruiting is delayed, additional insecticide applications may be necessary to protect small bolls that may be expected to mature.

Boll weevil. Cotton fields should be inspected weekly and when 15 to 25 percent of the squares are weevil-damaged from the time of squaring to peak flowering, the economic threshold level has been reached and insecticide application should be made. It may be necessary to repeat applications at five-day intervals or at three-day intervals if the weevil population increase is extremely great. If 60 percent or more of the bolls present after peak flowering are at least 3 cm in diameter, higher percentages of damaged squares can be tolerated. However, additional applications may be necessary to protect smaller bolls if they are to be harvested.

Bollworm and tobacco budworm. Tobacco budworm and bollworm moths are attracted to and lay eggs readily in cotton that is producing an abundance of new growth. Cotton is especially attractive and vulnerable to damage during the first four to six weeks of flowering. Moths usually lay eggs singly on the tops of young, tender, terminal leaves in the upper third of the plant. Eggs hatch in three to four days, turning light-brown before hatching. Young larvae usually feed for a day or two on tender leaves, leaf buds and small squares in the plant terminal before moving down the plant to attack larger squares and bolls. When small larvae are in the upper third of the plant, they are most vulnerable to control by insecticides, beneficial insects and spiders.

Tobacco budworm is less susceptible to certain insecticides than bollworm, but is less numerous than bollworm until mid-July. Once applications of insecticides are made to control bollworm and budworm, the percentage of budworm in the population increases with each additional application because of its greater tolerance to the chemicals.

Before flowering, the economic threshold is reached when larvae are present and 15 to 25 percent of the green squares are damaged.

Once bolls are present the economic threshold has been reached when larvae are present and 8 to 10 percent of the green squares have been damaged.

After the initiation of insecticide treatments, fields should be checked closely two or three days after the first application. The economic threshold has been reached when bollworm eggs and six to ten young larvae are found per 100 terminals (7 400 to 9 800 young larvae per hectare) and 5 percent of the squares and small bolls have been injured by small bollworm and budworm larvae. If control has not been obtained, another application should be made immediately.

EARLY HARVEST AND PHYTOSANITATION PRACTICES

Early harvest, stalk destruction and ploughing under of crop debris are among the most effective practices for reducing numbers of overwintering boll weevil and pink bollworm if carried out by all the farmers in the area.

Farmers are encouraged to mature their crop early so they can defoliate or desiccate the cotton by mid-September or earlier. In areas where it can be done, this practice alone may reduce numbers of overwintering boll weevil and pink bollworm by as much as 90 percent. This is because diapausing forms of the two pests occur during late summer and early autumn in response to shorter days, cooler temperatures and maturation of the cotton plant. In the case of the boll weevil, adults that emerge during this period feed on cotton bolls and squares for a few days then fly from cotton fields to nearby woodland where they overwinter in leaf litter. The pink bollworm diapauses as a last instar larva mainly in the seeds within dried-up bolls. When defoliation is followed by rapid harvesting, stalk shredding and ploughing under of debris after harvest, numbers of the pests are reduced even more. In many years these practices are so effective that the pests cause only negligible losses to cotton in the following season (Adkisson *et al.,* 1982).

To encourage these practices laws have been enacted in Texas, and are strictly enforced, that establish uniform early planting dates for cotton and dates when all cotton stalks must be shredded in the various cotton-producing areas of the state.

Whitefly

In the Lower Rio Grande Valley of Texas, whitefly has emerged as a serious secondary pest of cotton as a result of resistance to all insecticides. In this area the pest is being suppressed by a highly intensified management system involving:

- uniform early planting of smooth-leaved cotton varieties that are not preferred by whitefly;
- careful management of insecticides applied against other pests so as to preserve the natural enemies of the whitefly;
- early defoliation, harvest and stalk destruction of cotton;
- host-free period between cotton and vegetables;
- ploughing-under of infested vegetable debris after harvest;
- careful planning of time of production of vegetable hosts relative to the cotton season.

Early experience suggests that the severity of whitefly infestations on cotton and vegetable hosts may be greatly reduced by total crop ecosystem management practices if followed by all farmers in the region.

MANAGEMENT OF INSECTICIDE-RESISTANT STRAINS
OF BOLLWORMS AND TOBACCO BUDWORMS

In the United States strains of the bollworm and tobacco budworm (*Heliothis* spp.) have emerged that are resistant to all classes of synthetic insecticides; recently, for example, strains that are resistant to the pyrethroids have appeared. Pyrethroid insecticides are too valuable to cotton IPM programmes to be rendered ineffective by the development of resistant strains of *Heliothis* spp., however.

The cotton production system must be carefully managed to minimize damage from *Heliothis* spp. The goal of pyrethroid resistance management is to protect the crop, maintain profits and extend the useful life of this class of insecticide. In Texas resistance management tactics have been developed and implemented that have been successful in preserving the usefulness of the pyrethroids for *Heliothis* spp. control. The basic strategy in this approach is the use of insecticides other than the pyrethroids early in the production season. The use of the pyrethroids is restricted to the critical mid-season period when *Heliothis* spp. are likely to cause the most damage to the crop.

Recent research has shown that certain biological constraints are associated with the development of pyrethroid resistance in *Heliothis* spp. The reproductive rate of resistant females is less than half that of susceptible ones. Thus, early in the season, if pyrethroids are not used, the susceptible strain will virtually replace the pyrethroid-resistant one. For this reason, pyrethroid insecticides should not be used during the early season. If field monitoring indicates that numbers of pest insects are sufficient to cause damage to the crop, treatment should be restricted to an organophosphate or carbamate insecticide.

In the mid-season, when *Heliothis* spp. become prevalent, adults collected from pheromone traps are monitored for resistance in vials coated with a pyrethroid. When treatment becomes necessary, and if little or no resistance to the pyrethroids has been detected in the population, pyrethroids may be used for three or four applications over one generation of the pest. Once pyrethroid-resistant individuals are detected, treatments should be made with organophosphate or carbamate insecticides only. In addition, all measures should be taken to mature the crop as early as possible so that treatment is not extended for a long period. Thus, the growing of early-maturing varieties in a short-season production system is an effective way of managing insecticide resistance in *Heliothis* spp.

The use of these management procedures has successfully stabilized the level of resistance to pyrethroids by *Heliothis* spp. in Texas and the rest of the United States.

UTILIZING MICROBIAL INSECTICIDES IN COTTON IPM

Microbial products that are natural pathogens of the bollworm and the tobacco budworm are commercially available as preparations of *Bacillus thuringiensis* (*Bt*). Field studies indicate that microbials are best suited for square protection. They are slow-acting and should be used only against infestations of larvae during the squaring period and through the first ten days of flowering. They are not recommended for use after that period. Microbials are effective against infestations as great as 12 larvae per 100 plants (15 000 per hectare). They do not destroy predators and parasites, a characteristic that sets them apart from conventional insecticides. When numbers of predators and parasites are low, other insecticides provide more effective control.

Treatment of cotton fields with *Bt* should be restricted to those in which most of the larvae are not more than 0.6 cm long. Infestations of larger larvae should not be treated with *Bt*. Maximum effectiveness with *Bt* requires precise sampling of cotton plants during the fruiting period. Sampling should be conducted at least twice a week while squares are developing. Apply *Bt* with ground equipment at a volume application rate of 47 to 140 litres per hectare, or by air at a volume application rate of 19 to 47 litres per hectare. Good coverage is essential.

CONCLUSION

Most cotton farmers in Texas use some form of IPM for insect control. They have implemented the system to save cotton production from the ravages of insecticide resistant pest insects. In doing this they have increased yields, lowered production costs and increased profits. Insecticide use on cotton has been greatly reduced, by more than 50 percent from the high levels of the

1970s when control was dependent mostly on the use of chemicals (Adkisson *et. al.,* 1985).

IPM has made Texas cotton growers far more competitive economically and has led to a resurgence of the industry in the state. Texas now produces almost half of the cotton grown in the United States.

References

Adkisson, P.L., Frisbie, R.E., Thomas, J.G. & McWhorter, G.M. 1985. Impact of integrated pest management on several crops in the United States. *In* R.E. Frisbie & P.L. Adkisson, eds. Integrated pest management on major agricultural systems. *Texas Agric. Exp. Sta. Misc. Publ.,* 1616: 663-672.

Adkisson, P.L., Niles, G.A., Walker, J.K., Bird, L.S. & Scott, H.B. 1982. Controlling cotton's insect pests: a new system. *Sci.,* 26: 19-22.

Ahmad, Z. 1991. *Cotton pest control in Pakistan.* FAO Plant Protection Division. Rome, Italy. 76 pp. (Unpublished report)

Bell, A.A. 1984. Cotton production practices in the USA and world: Section B, diseases. *In* R.J. Kohel & C.F. Lewis, eds. *Cotton,* agronomy monograph No. 24, p. 288-309. Madison, WI, USA, American Society of Agronomy.

Bell, T.M. & Gillham, F.E.M. 1989. *The world of cotton.* Washington, DC, USA, Conticotton, EMR. 410 pp.

Chandler, J.M. 1984. Cotton production practices in the USA and world: Section D, weeds. *In* R.J. Kohel & C.F. Lewis, eds. *Cotton,* agronomy monograph No. 24, p. 330-365. Madison, WI, USA, American Society of Agronomy.

Elliot, F.C., Hoover, M. & Porter Jr., W.K. 1968. *Advances in production and utilization of cotton: principles and practices,* p. 3-4. Ames, IA, USA, Iowa State University Press.

FAO. 1972. *Report of the Fourth Session of the FAO Panel of Experts on Integrated Pest Control.* 6 to 9 December 1972. Rome, Italy. 35 pp.

FAO. 1975. *Pest management systems for the control of the pests of cotton,* a report on an FAO/UNEP consultation, 13 to 16 October 1975, Karachi, Pakistan. AGP 1976/M/3. Rome, Italy. 26 pp.

Frisbie, R.F., Hardee, D.D. & Wilson, L.T. 1992. Biologically intensive integrated pest management:future choices for cotton. *In* F.G. Zalom & W.E. Fry, eds. *Food, crop pests and the environment,* chapter 3. St. Paul, MN, USA, American Phytopathology Society.

Gulati, A.M. & Turner, J.A. 1928. *A note on the early history of cotton.* Indian Cent. Cot. Comm. Bulletin No. 17 Technical Service No. 12.

Kohel, R.J. & Lewis, C.F., eds. 1984. *Cotton,* agronomy monograph No. 24. Madison, WI, USA, American Society of Agronomy. 605 pp.

Norman Jr., J.W. & Sparks Jr., A. 1991. Management of cotton insects in the Lower Rio Grande Valley of Texas, 1991-1992. *Tex. Agric. Ext. Serv. Bull.,* 1210. 10 pp.

Parker, R.D., Huffman, R.L. & Sansone, C.G. 1991. Management of cotton insects in the southern, eastern and Blackland areas of Texas, 1991-1992. *Tex. Agric. Ext. Serv. Bull.,* 1204. 11 pp.

Pearson, E.O. & Darling, R.C.M. 1958. *The insect pests of cotton in tropical Africa.* London, UK, Empire Cotton Grow. Corp. and Commonw. Inst. Entomol. 355 pp.

Reynolds, H.T., Adkisson, P.L., Smith, R.F. & Frisbie, R.E. 1975. Cotton insect pest management. *In* R.L. Metcalf & W. Luckmann, eds. *Introduction to insect pest management.* p. 379-443. New York, NY, USA, John Wiley and Sons.

Smith, R.F. & Reynolds, H.T. 1972. Effects of manipulation of the cotton agro-ecosystem on insect pest populations. *In* M.T. Farvar & J.P. Milton, eds. *The careless technology – ecology and international development,* p. 373-406. Garden City, NY, USA, Natural History Press.

USDA. 1992. *World cotton situation.* United States Department of Agriculture, Foreign Agricultural Service, Circ. Serv.FC 9-92, September 1992. Washington, DC, USA. 36 pp.

Interface and application of biotechnology to cotton improvement

Kamal M. El-Zik

ABSTRACT

Cotton breeders have made major contributions to the increases in yield, fibre quality and resistance to pests of the past 20 years. Traditional plant breeding involves making large numbers of crosses between a range of parents which have been selected by the breeder for their desirable attributes. Crosses are made among selected parents to increase the genetic variability by transferring, combining and pyramiding genes. Some of the progenies from the crosses, after selection and testing, show a combination of the best traits from the parents. It takes 12 years from the time a cross is made to the selection and release of a cultivar. The breeding and genetic improvement process is simple in outline but complex in practice and requires a wide range of skills and a basic knowledge of several different sciences. Successful programmes require a team effort from many disciplines.

New tools are needed in order to enhance efforts in developing superior cotton cultivars for the year 2000 and beyond. Plant breeding programmes are a long-term commitment, they require the growing and analysis of large populations, which take both time and space. In addition, in some cases breeders have limited genetic variation for use in developing new cultivars.

Recent technical advances in recombinant DNA and tissue culture methods have developed to a point where biotechnology will have a major impact on the efficiency and success of cotton breeding. Biotechnology and genetic engineering assist breeders to improve crops by increasing the efficiency and effectiveness of selection, broadening the genetic base from which breeders can select through direct or indirect genetic modification, introducting new cotton genes and introducing novel genes.

There are five main components to cotton biotechnology: DNA marker development – the gene must be available as a discrete entity of DNA; identification and isolation of agriculturally important genes; cotton transformation; development of stable and safe transgenic plants; and release of new cotton cultivars (intellectual property issues). These components must all be present for the successful integration and development of transgenic plants.

Why are markers needed? The first step in developing genetic maps is to track and follow the gene through a cross and identify its position on the chromosome. When genes of interest have markers, recombinations between these genes can be detected and accurate selection for genetically superior individuals, from among the large number of segregating populations becomes easier. This is the tool – F_2 – that allows selection for genotypes carrying the specific gene(s) to take place at the seedling stage or at an early generation rather than waiting to identify the phenotype in the field by marker assisted selection. Identifying markers linked to agriculturally important genes is the first necessary step for gene cloning and allows for transfer of specific genes to elite cotton lines or superior cultivars.

Which genes are of critical importance to cotton productivity and need to be identified and isolated? Genes that control resistance to the insects and pathogens that cause diseases, resistance or tolerance to herbicides, physiological stresses, physiological processes and quality and productivity of the cotton crop are all of economic importance. It is in the area of crop protection that the recent advances in biotechnology will provide new solutions.

There are prerequisites for gene transformation techniques. Genetic transformation can be defined as the transfer of foreign genes isolated from plants, viruses, bacteria or animals into a new genetic background. The gene must be available as a discrete entity of DNA. It has to be manipulated in a transfer system for which stable transformants must be generated. The cell must have the capacity to differentiate into a stable and safe transgenic plant. In plants, successful genetic transformation requires the production of normal-fertile plants that express the newly inserted genes. The process of genetic transformation involves several distinct stages: insertion, integration, expression and inheritance of the new DNA. It is still early days in the development of biotechnology and most of the work to date has concentrated on the introduction of a single gene that confers the desired resistance character.

Regeneration of transgenic cotton plants is an obstacle to the full utilization of biotechnology. The *Bacillus thuringiensis* and herbicide tolerance genes have been introduced in only a limited number of cotton cultivars, mainly Coker 312 and its sister lines. New techniques for regeneration of cotton plants are needed. Biotechnology can never replace or be used independently from conventional cotton breeding. Biotechnology does not replace existing science but rather adds new tools for use.

Genetic enhancement and improvement of crops are based on creating genetic variability, followed by selection. Variability, created normally by conventional breeding through sexual recombination, may be increased substantially by biotechnology. Biotechnology can assist improvement of the crop by broadening the base from which the breeder can select through direct or indirect genetic manipulation, thus increasing the efficiency and effectiveness of selection. Cotton breeding will continue to deal with complex characters, the inheritance of which follows the laws of population genetics. The combination of several quantitatively inherited characters still demands enormous populations and quantitative selection procedures. The development of computer software and systems analysis to aid breeders and biotechnologists in their diverse areas of science is critical.

The short-term products of biotechnology are new and better solutions to problems that scientists have been working to solve for many years by conventional technology, for example, high resistance to insects and diseases, herbicide resistance, drought tolerance and improved fibre quality. Biotechnology offers new basic knowledge, techniques and products for the future. Bridging and merging biotechnology, molecular biology and cotton breeding is essential for the twenty-first century.

Part III
Consultation report

Round-table discussions

In this section, the main points to emerge from the round-table discussions are summarized.

ECONOMIC IMPORTANCE OF COTTON IN THE REGION

In all the participating countries cotton is a major cash crop and an important source of export earnings, from raw cotton, yarn or piece goods. In Egypt, Pakistan, the Sudan and the Syrian Arab Republic, cotton is central to the national economy. Cotton production provides cash income for farmers and rural and urban employment in transport, ginning, processing and textile manufacturing. Cottonseed oil for domestic consumption and cottonseed cake for animal feed are important by-products of the cotton industry.

The weather and insect pests are the most important factors determining cotton yields in most countries. Because of the economic importance of cotton, governments provide support to farmers to reduce the risks that pests represent to the individual farmer and the economy as a whole. Support takes various forms, including technical assistance, advice and training on pest management, pest monitoring and forecasting services, control operations and credit and subsidies for pesticides and other inputs required by the cotton crop. In spite of a high level of government support in most countries in the region, losses to pests remain high and the costs of these losses, when combined with the costs of control, amount to many hundreds of millions of United States dollars each year.

COTTON PRODUCTION

Table 43 gives the cotton production statistics for the countries of the region participating in the consultation.

CULTURAL PRACTICES AND VARIETIES

Cotton in the Near East Region is grown between the latitudes 10° north (the Sudan) and 40° north (Turkey), in climates that are mainly hot and dry. Most

TABLE 43
Cotton area, production and yield for the Near East region

County	Year	Area (ha)	Lint production (tonnes)	Lint yield (kg per ha)
Egypt	1993	371 000	358 000	906
Iran	1993	205 000	114 000	556
Iraq	1991	7 000	5 000	297
Morocco	1991	7 500	4 600	613
Pakistan	1993	2 893 000	1 348 950	481
Sudan	1993	159 600	67 642	423
Syria	1993	196 000	223 000	1 140
Turkey	1991	599 000	559 000	935
Yemen	1993	18 700	12 622	675

cotton is therefore grown under irrigation although some is rain-grown. Water management is an important factor in limiting the impact of pests, too much water or prolonging watering into the harvest period may create a favourable environment for pests and facilitate the carryover of populations from one season to the next. The overuse of nitrogen may similarly create favourable conditions for the buildup of pests.

The main varieties of cotton grown in the region are shown in Table 44. Nearly all production is of the New World species *Gossypium hirsutum* and *G. barbadense*. Cottons of Asiatic origin are now rarely grown. Most of the production is from medium-staple American upland varieties, although longer-staple Acala types are grown in several countries. Long-staple, Egyptian (*G. barbadense*) varieties are mainly grown in Egypt and the Sudan, with some production in other countries. Most of the varieties grown require a growing season of at least 180 to 200 days, although some shorter-season varieties are grown at higher latitudes.

ARTHROPOD PESTS
The main insect and mite pests occurring in the region are listed in Table 45 where they are divided into early-season pests and mid- to late-season pests,

TABLE 44

Principal cotton varieties currently grown in the Near East region

Region	Variety	
	Gossypium hirsutum	*Gossypium barbadense*
Egyt		Dandarah Giza 45, 70, 75, 76, 77, 80, 81, 85
Iran	Bakhtegan, Oltan, Pake, Sahel, Varamin, Hopicala, Deltapine 16, Coker 100 Wilt	
Iraq	Coker 310	
Morocco		Pima 67, Tadla 16
Pakistan	MNH-93, NIAB-78, MS 84, SLH-41, CIM 240, CIM-109, S-12, RH-1, Gohar 87, K-68-9, Rehmani, Shaheen, BH 36, MNH-147	
Sudan	Acala, Shambat B	Barakat 90
Syria	Aleppo 40, Aleppo 33/1, Rakka 5, Deir Ezzore	
Turkey	Cukurova 1518, Nazilli 84, Nazilli 87, Del Cero, Sayar 314, Marasa 92, McNair 235, Deltapine 20, 15/21, 61, Ersan 92, Stoneville 825	
Yemen	Coker 100 Wilt, Acala Sj2	K4

with their status in each country shown. Most of the pests occur in most of the countries but there are variations in their relative economic importance. Early-season pests are mainly sucking pests, although in Yemen early-season termite attack may cause considerable loss of stand, as do attacks on seedlings by cutworms in a number of countries. Many sucking pests are also mid- to late-season pests. These reduce yield indirectly by feeding on leaves, stems and growing points (sucking pests and leafworms) or directly by feeding on the reproductive parts of the plant (buds, flowers and bolls). The latter group includes bollworms and various bugs, such as the shedder bug. A third group of mid- to late-season pests reduces lint quality by excreting honeydew which contaminates the lint, causing stickiness and promoting the growth of moulds. Aphid and whitefly are the main species involved.

The most important pests, taking the region as a whole, are pink bollworm and *Earias* spp. (spiny and spotted bollworms), together with the sucking pest complex, the components of which vary from country to country but

TABLE 45

Arthropod pests of cotton reported in the Near East region

	Egypt	Iran	Iraq	Morocco	Pakistan	Sudan	Syria	Turkey	Yemen
Early-season									
Cutworms									
(*Agrotis* spp.)	+						+	+	
Thrips									
(*Thrips tabaci*)	+	+	+		+		+	+	+
Thrips									
(*Scirtothrips dorsalis*)					+				
Spider mites (various spp.)	++	+	+++	+					
Flea beetle									
(*Podagrica* spp.)						+		+	+
Aphid									
(*Aphis gossypii*)	+	+	+	++			+	++	
Mid- and late-season									
American bollworm									
(*Helicoverpa armigera*)	+	+++		+++	++	++	+++	+++	++
Pink bollworm									
(*Pectinophora gossypiella*)	+++	+		+++	+++	+		+	++
Spiny bollworm									
(*Earias insulana*)	++	+++	+++	+++	+++	+	+++	+	++
Spotted bollworm									
(*Earias vittella*)					+++				
Red bollworm									
(*Diparopsis watersi*)						+			+++
Egyptian cotton leafworm									
(*Spodoptera littoralis*)	+++	++		+				++	++
Cotton leafworm									
(*Spodoptera litura*)					+				
Lesser armyworm									
(*Spodoptera exigua*)	+	+					+	+	+
Whitefly									
(*Bemisia tabaci*)	++	+++	++	+++	+++	+++	+	+++	++
Cotton jassid (various spp.)	+				++	++	+	++	
Spider mites (various spp.)		++	+++		+			++	++
Aphid									
(*Aphis gossypii*)	+	+	+++	+	+++		+++	++	
Cotton stainer									
(*Dysdercus spp.*)					+	+			
Termite (various spp.)					+				++
Shedder bug									
(*Creontiades pallidus*)							++	+	

+++ = major pest; ++ = moderate pest; + = minor pest.

commonly include aphid and whitefly. The importance of the sucking pest complex may be caused by the direct effect it has on yield when infestations are high, to the reduction in quality it causes with its stickiness or because it is a vector of diseases, notably cotton leaf curl virus in Pakistan.

The status of particular pests in many countries has been affected by the use of pesticides on other pests in the complex. Thus pests which in the absence of pesticides would be kept at low levels by natural enemies, have become important or potentially important, as pesticide use has reduced the effectiveness of natural control agents. Examples include whitefly and American bollworm. It is noteworthy that both these species have become highly resistant to insecticides in a number of countries. A few pests, notably the cotton leafworm in Egypt, are of considerable importance in one country, but of only minor significance in neighbouring countries.

When interventions are required, conventional insecticides and acaricides remain the most common method of pest control on cotton in all the countries of the region. However in all countries a variety of measures are employed or recommended to keep pest populations below economic threshold levels so that chemical control interventions are not needed. These measures include varietal resistance and various cultural practices, often backed up by legislation. In some countries alternatives to conventional chemical pesticides, such as pheromones and microbial insecticides, are being employed or are the subject of research.

Alternative control methods are becoming increasingly important as resistance to many conventional pesticides becomes more of a problem. In much of the region whitefly, aphid, American bollworm and leafworm have become tolerant or resistant to many organophosphate, carbamate and pyrethroid insecticides, leading to some countries reporting great difficulty in controlling these pests. Other pests, notably pink bollworm, do not as yet appear to have developed resistance. In some countries resistance-management strategies have been adopted, involving the alternation of different chemical groups from one spray to the next, the restriction of the use of pyrethroids to the mid- and late season, the use of microbial insecticides, particularly *Bt* to control Lepidopterous pests early in the season and the prohibition of the use of insecticides against whitefly. In spite of these measures pests continue to develop resistance which now constitutes a major threat to the sustainability of cotton production in the region.

DISEASES

The major diseases of cotton occurring in the region are listed in Table 46. All the countries in the region report problems with seedling diseases, including damping-off, root rot, leaf curl, bacterial blight and anthracnose, and measures are taken in all countries to reduce the impact of these diseases. These measures include:

- use of good quality seed;
- use of varieties resistant to bacterial blight;
- high seed rates;
- sulphuric acid treatment of seed;
- fungicidal seed dressings;
- field sanitation and destruction of crop residues.

Verticillium wilt is serious in the Islamic Republic of Iran and Syria, while Iran also reported fusarium wilt as a problem. Wilt tolerant or resistant

TABLE 46

Diseases of cotton in the Near East region

	Egypt	Iran	Iraq	Morocco	Pakistan	Sudan	Syria	Turkey	Yemen
Bacterial blight (*Xanthomonas campestris*)				+	+++	++		+	+
Cotton leaf curl virus (CLCV)				+	+++	+			
Damping-off (*Rhizoctonia solani*)	+	++	+++	+	++	+	+	++	++
Fusarium wilt (*Fusarium oxysporum*)		++		+					
Veticillium wilt (*Verticillium dahliae*)		+++	+++	+			+++	++	
Anthracnose	+				+				
Leaf spot (*Alternaria* spp.)				+					+
Boll rots (*Aspergillus flavus, Diplodia gossypina, Colletotrichum capsici, Xanthomonas campestris, Myrothecium roridum*)				+	++				++

+++ = major disease; ++ = moderate disease; + = minor disease.

varieties are the best solution to the wilt problem, which can become exacerbated by the overuse of nitrogenous fertilizer. Cotton leaf curl virus has been a serious problem in Pakistan in recent years; varietal resistance would appear to be the long-term solution.

Boll rots are a problem, especially in irrigated cotton and where bolls have already been damaged by insects. Effective pest control and careful water and fertilizer use to avoid rank growth and high humidity during boll maturation, can limit the importance of boll rots.

WEEDS

The major weeds species are listed in Table 47. The perennial grass weeds *Cynodon dactylon* and *Sorghum halepense,* the annual broad-leaved weeds *Solanum nigrum, Amaranthus* sp. and the perennial sedge *Cyperus rotundus* are the main problems. Mechanical and hand-weeding are the main

TABLE 47

Weeds in cotton in the Near East region

	Egypt	Iran	Iraq	Morocco	Pakistan	Sudan	Syria	Turkey	Yemen
Cynodon dactylon				+++	++	+++	+++		+++
Xanthium spp.	+		+	+			+	+++	
Convolvulus arvensis				++	+		++		
Setaria sap							++	++	
Sorghum halepense			+	+	+++		+++	+++	
Chenopodium spp.	+	++		+			+		
Solanum spp.		++	+	+++			+	++	++
Portuluca oleracea	+		+	+			+		
Panicum colonum							+		
Amaranthus spp.	++	++	++	++	+		++	++	++
Cyperus rotundus				++	++	++		++	+++
Ischaemum afrum						++			
Trianthema monogyna					+++				
Brachiaria eruciformis				+					
Tribulus terrestris					+				
Digitaria spp.				+					
Hibiscus trionum	+	+							
Malva parviflora	+	+							
Corchorus olitorius	+								
Echinochloa spp.	+	+	+++						++
Euphorbia spp.	+	++			+				+
Polygonum aviculare			+++						

+++ = major pest; ++ = moderate pest; + = minor pest.

methods of control throughout the region. Hand-weeding provides employment during the growing season for people who are later needed to harvest the crop. Some use of pre-emergence herbicides occurs in most countries and similarly there is some use of post-emergence herbicides for spot treatments. In a few countries herbicides are becoming the main method of weed control in cotton.

INFRASTRUCTURAL SUPPORT
Research

Each of the participating countries at the consultation has research programmes on cotton pests, diseases and weeds. The ministry of agriculture or its equivalent normally takes the lead in this research through its specialized crop protection research institutes or in multidisciplinary research institutes. In some countries cotton research is the responsibility of cotton commodity organizations. In most countries university-based research makes an important contribution. Aid organizations and donor countries, including the United Nations Development Programme (UNDP), FAO, the United States Agency for International Development (USAID), the Overseas Development Administration of the United Kingdom (ODA/UK) and the Government of the Netherlands, have provided funding and technical assistance for cotton pest research in many countries. The major cotton producing countries, namely Egypt, Pakistan, Turkey and the Sudan, have the largest research programmes to support their production. While there is a considerable number of experienced and competent cotton scientists in the region, research is constrained by a number of factors including:

- inadequate funding for research projects;
- funding instability;
- rising costs of personnel, facilities and operating expenses;
- lack of modern research facilities and equipment;
- need for overseas training at the postgraduate level (M.Sc. and Ph.D.) for cotton scientists.

Extension
There are extension services in all the participating countries but constraints, especially with regard to funding, reduce their effectiveness in many places. Research-extension links are often weak, which further reduces effectiveness.

Plant protection services and quarantine
Each country has plant protection and plant quarantine services although their role in relation to cotton pest management varies from country to country. The services may be responsible for pest forecasting and monitoring, for the provision of advice to farmers and for operational pest control as well as the enforcement of legislation covering pest control. Often staff of these organizations have more contact with farmers in connection with cotton than the extension services do. Resource constraints again limit the effectiveness of these organizations in many countries and in several the enforcement of legislation is not rigorous and needs strengthening.

A BASIC STRATEGY FOR IPM
The first requirement for developing an integrated pest management (IPM) strategy for cotton is an understanding of what is meant by IPM. The FAO Panel of Experts on Integrated Pest Control (1972) defined integrated pest control as "A pest management system that, in the context of the associated environment and the population dynamics of the pest species, utilizes all suitable techniques and methods in as compatible a manner as possible and maintains pest populations at levels below those causing economic injury". This broad definition implies the fullest use of natural mortality factors complemented when necessary by artificial methods. Also implicit in this definition is the belief that chemical control methods should be used only when economic injury thresholds would otherwise be exceeded and, when used at all, selective pesticides should be preferred over those with a broad spectrum of activity and should be applied in a manner that maximizes their effectiveness and minimizes harm to the environment.

Arthropod pests

The basic strategy of an IPM programme for the arthropod pests of cotton is simple; use tactics that suppress the key pests while conserving the natural enemies of key and major secondary pests. Basic objectives are to avoid key pest resurgence and secondary pest outbreaks. The use of insecticides must be carefully managed so as to avoid the destruction of parasitoids and predators. Alternative, non-chemical methods of pest suppression and natural enemy conservation must be employed as broadly as possible to minimize or eliminate the need for treating cotton fields with chemical insecticides. Whatever tactic is chosen, its harmonized implementation by the majority of farmers in a community will have a cumulative effect on the suppression of the target pest species.

In developing the tactics for implementing this strategy, the first consideration should be the development and adoption by farmers of crop and habitat management practices and cultural control procedures that suppress the pests while conserving their natural enemies. For most pests, insecticides should be used only as a last resort and spraying should be based on the use of economic injury thresholds for the pests concerned.

Crop management practices. These should include the following:
- Crops should be rotated and cotton fields deep-ploughed to reduce numbers of soil pests and overwintering or diapause stages of pink and American bollworms, leaf- and armyworms. Fields should be irrigated after ploughing to reduce further the pest numbers by stimulating the suicidal emergence of moths during host-free periods.
- Seed for sowing should be treated to kill diapause pink bollworm larvae and seedling pests.
- Seed should be of good quality, of pest resistant or tolerant varieties and it should be sown at a rate sufficient to produce a good stand of vigorous plants.
- In areas infested by whitefly, growing alternative vegetable hosts near cotton fields should be avoided. Vegetables should be grown at enough distance from cotton to prevent migration of whitefly on to cotton.
- Overuse of nitrogenous fertilizers and irrigation water should be avoided.

Excessive or prolonged irrigation during the boll-opening period can greatly increase the severity of whitefly infestations.

• Weeds should be controlled, especially those that are hosts of whitefly, spiny, spotted and red bollworms.

• The harvest period should be as short as possible and should be followed by the grazing of fields by stock and the disposal of crop residues by methods that reduce pest numbers, especially pink bollworm. Chemical defoliants should be used if whitefly or aphid are serious problems.

• Residues of vegetable hosts of whitefly should be ploughed in immediately after harvest.

• Crop development should be monitored and the occurrence of different pests species related to crop phenology.

Legislative measures. All participating countries have promulgated legislative measures for the control of the insect pests of cotton. Measures that need to be taken are, for the most part, included in the regulations but enforcement is often not sufficiently rigorous to be effective. In general, measures included in regulations must be carried out by all farmers over large areas to be effective in controlling pest populations. The following are some of the regulatory measures adopted:

• Mandatory destruction of cotton residues after harvest, including the pulling up, stacking and burning of stalks and dry bolls, is the most common and effective method for controlling the pink bollworm.

• A cotton-free period between harvest and the following season's cotton crop is enforced.

• Cottonseed is treated to kill diapausing pink bollworm larvae. There are procedures for handling raw cotton and cottonseed at ginneries and farmers' homes to kill diapause pink bollworm larvae during the non-cotton season. The storage as winter fuel of cotton stalks that bear dried bolls, which may be infested with pink bollworm is prohibited.

• Uniform sowing dates for cotton are enforced.

• Alternate hosts of cotton pests are treated so as to minimize pest infestations on cotton (e.g. irrigation of clover in Egypt to reduce cotton leafworm numbers, prohibition of okra cultivation during certain periods

of the cotton season to reduce infestations of cotton by pink and spiny bollworm).
- Leafworm egg masses are collected by hand (for example in Egypt).
- The use of insecticides is regulated (implementation of the FAO International Code of Conduct on the Distribution and Use of Pesticides).

Control with insecticides. Insecticides are a powerful weapon for the control of the arthropod pests of cotton. As such, they should be used only when needed and at the minimum dose required for control. Except for seed treatment, applications should be based on economic injury levels and frequent inspection of cotton fields. Every attempt should be made to conserve natural enemies, to avoid target pest resurgence and secondary pest outbreaks and to minimize the development of insecticide resistant pest strains. As far as possible, selective insecticides are to be preferred to broad spectrum ones. Measures that have been employed to achieve these objectives include:
- Economic threshold levels for several pest species have recently been greatly increased, thereby delaying the application of insecticides on cotton by days or weeks. This gives indigenous natural enemies time to increase in numbers to such an extent that spraying can sometimes be entirely avoided.
- Cotton fields are checked frequently for pests and treatments are applied only as needed. This avoids large-scale broadcast treatments of entire cotton-growing areas.
- Low dosages of less harsh and more selective insecticides are used to control early-season pests. This conserves the natural enemies of mid- and late-season pests.
- Some Lepidopterous pests may be controlled either by applying *Bt* or by disrupting mating using pheromones, conserving natural enemies at the same time.
- Insecticidal treatments on cotton can be alternated between organophosphate, carbamate and pyrethroid insecticides in an attempt to manage the development of insecticide resistant pest strains. The use of pyrethroid insecticides can be restricted to a certain period during the growing season or to a few applications.

- The use of insecticides against whitefly may be prohibited in certain countries.
- The determination of the need for treatment with insecticides, the choice of chemicals and spraying operations may be under the control of plant protection specialists employed by the government, but farmers should participate in the decision-making process.
- Pesticides for use on cotton should not be subsidized by governments, as this practice stimulates pesticide overuse.
- Safe and efficient use of pesticides should be promoted.

Conclusions. The basic elements needed for the development of IPM for arthropod pests on cotton in the Near East are known. Successful IPM programmes have been implemented in some countries in the region and could be implemented in the others. Research will continue to improve and develop the tactics required. There is a need to develop multiple diversity resistant cultivars integrating resistance to insect pests with resistance to other crop production constraints (including pathogens, drought, high or low temperatures and soil salinity). There is also an urgent need for more research to prevent or delay the further development of insecticide resistant pest strains. If this is not done, some pests, especially whitefly, American bollworm and cotton aphid, may develop strains resistant to all insecticides. As present research is fragmented; efforts must be made to make it multidisciplinary and farmer-driven. The implementation of IPM on cotton is dependent on effective and highly motivated extension and plant protection services and these must be strengthened in all countries if IPM is to be successful.

Finally, government policies that encourage the overuse of pesticides, through subsidies for example, need to be re-examined. Successful IPM depends on the rational use of pesticides.

Plant diseases

The basic strategy for controlling diseases of the cotton plant is to reduce the pathogen inoculum or to restrict development of the pathogens in the tissues of the host plant. This strategy is best achieved through the integration of several tactics, including plant sanitation, cultural control, the growing of disease resistant varieties and the use of seed dressings.

Crop management practices. Practices that may be adopted to control or reduce the severity of diseases of cotton include sowing, fertilization, irrigation, crop rotation, tillage and phytosanitary practices. They should include the following:

- delay of cotton sowing until the soil temperature is 18°C or higher to ensure rapid germination of the seed;
- sowing of high-quality seed at rates and depths sufficient to provide good stands of vigorous plants – this is important to reduce the severity of seedling diseases as well as some late-season diseases such as verticillium wilt;
- sowing seed of varieties resistant to bacterial blight, fusarium wilt and other diseases;
- treating seed with an effective fungicide;
- applying fertilizer, especially potassium, at optimum levels for plant growth and yield; avoidance of overfertilization with nitrogen;
- avoidance of inadequate or excessive irrigation during the main growing season – restricting irrigation to a minimum in the late season when temperatures begin to fall;
- Controlling boll-feeding insect pests to reduce the incidence of boll rots;
- using clean tillage practices after harvest (grazing, ploughing and burning), completely destroying or burying cotton debris to prevent an increase in inoculum and early-season infections from seedling and foliar pathogens and viruses;
- rotating cotton with small grains, sorghum, legumes or paddy rice.

Conclusions. These crop management practices will give control or greatly reduce the severity of most of the diseases that infect cotton plants. Diseases cause considerable reductions in the yield of cotton and more effective control measures are still needed. Such methods will come only through research to develop new disease resistant cultivars, enhance biological control and produce more effective and economical chemical controls. Additional funding is needed in these research areas.

Weeds

The most effective way to control weeds in cotton is through the integration of cultural, mechanical and chemical control measures. Measures that may be used include:

- use of manual labour to hoe, cut or remove weeds from cotton fields and adjacent areas;
- ploughing-in of weeds when preparing the land for sowing, followed by hoeing during the growing season to destroy weeds growing between the cotton rows;
- use of pre-emergence herbicides to control weeds in the early- to mid-season period;
- selective use of post-emergence herbicides to control specific weed species, including spot treatment to kill individual plants or small clumps of weeds – broadcast treatments may be used where weed infestations are severe;
- rotation of cotton with other crops to permit the use of additional tillage practices – different types of herbicides may also be used to control particularly difficult weed infestations;
- fallowing of land, combined with tillage, may also be used to control dense populations of weeds.

The biological control of weed species infesting cotton has not been effective and will require a great deal more research before becoming practical.

Conclusions. Effective methods for controlling the weeds of cotton are known and used in the cotton-producing areas of the Near East. An abundance of labour allows weed control to be carried out mainly by human labour and ploughing. Herbicides are used and could be used on a much larger scale, but this might lead to problems with the displacement of farm labour, environmental degradation and sustainability. Research is needed to develop more effective IPM programmes for weeds, taking into consideration the socio-economic issues involved. Specifically, new research should be directed towards the development of intercropping and the use of cover crops to suppress weeds, and these could be linked with conservation tillage systems. Planting patterns and planting geometry should be investigated to determine optimum weed suppression.

GENERAL CONCLUSIONS AND RECOMMENDATIONS

There is a sufficient knowledge base and cadre of specialists in the Near East to formulate an IPM system for cotton for the region. This does not mean that all the necessary research has been done, nor that the extension and plant protection services have sufficient resources to provide the technical assistance to farmers that is required to implement IPM on a large scale. Research, extension and regulatory services will need to be greatly strengthened if a regional IPM programme is to be successfully implemented and sustained.

The consultation discussed ways and means to develop and implement such programmes and endorsed the following recommendations addressed to governments:

- Cotton producing countries should revise, enhance and strengthen their national cotton programmes in a way that enables the development and implementation of strategies for integrated pest management on cotton.
- The donor community and FAO should be approached with requests for the formulation, funding and implementation of a regional integrated pest management project based on the proposal given on p. 265.

Adoption of report

After the introduction of minor changes, the report of the consultation was unanimously adopted.

Closing session

Participants expressed their gratitude to the consultation organizers and hosts and expressed their satisfaction with the outcome of the consultation. Mr Barbosa and Mr Taher thanked the participants for their fruitful contributions which led to the achievement of the consultation objectives. They expressed the hope that follow-up actions on the recommendations of the consultation would be initiated at both national and regional levels.They concluded by expressing sincere thanks to the host country for the hospitality and the excellent organization of the meeting. Mr Saydam expressed his satisfaction with the results of the meeting and hoped that this consultation would bring about more cooperation and coordination among the countries of the Near East.

Proposal for a regional integrated pest management project on cotton for the Near East

INTRODUCTION

Cotton is a major cash and export crop in the Near East. The area given over to cotton production exceeds 4.5 million ha. Lint is exported for hard currency and used domestically to supply clothing needs. Cottonseed is used as a primary source of cooking oil and livestock feed. The growing and processing of cotton provides employment for millions of people in the Near East. It is therefore very important that cotton production should be sustainable in this region of the world.

Many arthropod, disease and weed pests are constraints to cotton production in the Near East. Of these pests, insects are the most serious and are often the most important factor limiting yield. In order to protect the crop from insect and mite pests, large quantities of chemical pesticides are used on cotton each year. The heavy use of these insecticides has not been without problems. They are not only expensive to purchase and apply but their use has led to target pest resurgence, secondary pest outbreaks, development of pesticide resistant pest strains, poisoning of workers and contamination of the environment. In several areas, control of certain pests (e.g. whitefly) by insecticides is no longer sustainable. For all these reasons, there is a need to develop and implement IPM programmes for cotton in the Near East.

The basic elements for IPM for pests of cotton in this region have been developed and utilized in several countries. The governments and plant protection institutes of the region should be encouraged to become more involved in research and training programmes on cotton IPM. Donor countries and international agencies should be invited to provide additional financial support and technical assistance. Host countries should be committed to regulating and managing the use of pesticides in a manner compatible with IPM. If all of these elements are adopted, IPM could be successfully implemented across the cotton-growing areas of the region.

The initiation of a regional IPM programme for cotton could focus a wide range of expertise in research and training on the problem. It could also provide a mechanism for coordination and collaboration across the region and between this region and other cotton-producing areas of the world.

Considerable effort has been made, and continues to be made, on the development and implementation of IPM for cotton in the Near East region. Much of the research needed to develop IPM for cotton has been done by national programmes with FAO and donor support. This needs to be continued and expanded. Greater effort is now needed to demonstrate and implement results and to strengthen the enforcement of uniform planting dates and other cultural practices. There also needs to be a revision of those government policies that encourage the overuse of insecticides on cotton. IPM cannot be successful unless pesticides are used on a rational basis.

DEVELOPMENT OBJECTIVES

The major development objective of the proposed project is to increase cotton production in the Near East in a sustainable manner, with reduced dependence on pesticides.

Immediate objectives

Policies favouring cotton IPM should be developed and implemented by national governments to generate the following outputs:

- policy statements on cotton IPM issued at national and provincial levels;
- policy mechanisms and directives;
- regulatory measures.

Cooperation and exchange of information and experience among participating countries should be promoted with the following outputs in mind:

- a national and regional coordination network for cotton IPM;
- a regional database on cotton protection;
- a newsletter on cotton IPM;

• coordination with cotton IPM programmes outside the region.

Farmer-driven research is required to enhance the development of cotton IPM by producing:

• regional programmes on implementation of research projects that are required for development of cotton IPM packages;
• IPM technologies at farmer level;
• evaluated research results.

Training programmes on cotton IPM practices need to be developed and implemented in order to achieve the following:

• policy-makers trained in IPM concepts;
• extension agents trained to disseminate IPM technologies;
• farmers trained in cotton IPM practices.

The national infrastructures involved in cotton IPM need to be strengthened in order to establish the following ends:

• linkages among all organizations concerned with cotton production and protection;
• plant protection and quarantine staff oriented towards the cotton IPM programme;
• revised and enforced plant quarantine regulations;
• the resources required for implementation of the programme secured.

Annexes

**1. PROGRAMME OF THE FAO EXPERT CONSULTATION ON COTTON PESTS AND THEIR CONTROL IN THE NEAR EAST REGION
5 to 9 September 1994, Izmir, Turkey**

Monday 5 September 1994
Morning session
- Registration
- Opening session
- Cotton production and protection with special reference to the Mediterranean-Near East region – Mr Michel Braud, France

Afternoon session
- Cotton pests and their control in Egypt – Mr Gallal Moawad
- Cotton pests and their control in Iran – Mr Ahmad Rassipour
- Cotton pests and their control in Morocco – Mr L. El-Jadd
- Discussion

Tuesday 6 September 1994
Morning session
- Important cotton diseases in the Near East: challenges and solutions – Mr Kamal El-Zik, United States
- Cotton pests and their control in Pakistan – Mr Zahoor Ahmad
- Cotton pests and their control in the Sudan – Mr N. Sharaf El-Din
- Discussion

Afternoon session
- Cotton pests and their control in Syria – Mr Farid Khouri
- Cotton pests and their control in Turkey – Mr Saban Karaat
- Cotton pests and their control in Yemen – Mr Saeed Ba-Angood

- Integrated pest management of cotton insects in Texas, United States – Mr Perry Adkisson (presented by Mr Sebastian Barbosa)
- Interface and application of biotechnology to cotton improvement – Mr Kamal M. El-Zik
- Discussion

Wednesday 7 September 1994
- Field trip

Thursday 8 September 1994
- Round-table discussions
- Conclusions and recommendations

Friday 9 September 1994
- Adoption of report
- Closing session

2. PARTICIPANTS

Participating countries

Egypt

Moawad, Gallal Mahmoud
Director
Plant Protection Research Institute
Agricultural Research Centre
Dokki, Giza, Cairo
Tel: 702193, 3486163
Fax: 716175, 716176

Morocco

El Jadd, Lahoucine
Head, Cotton Programme and
Research Centre
National Institute for Agronomic
Research
Beni-Mellal
Tel: 440006
Fax: 440083

Pakistan

Ahmad, Zahoor
Director
Central Cotton Research Institute
PO Box 572
Old Shuja Abad Road
Multan
Tel: 30151/2
Fax: 44153

Sudan

Sharaf Eldin, N.
Agricultural Research Corporation
PO Box 126
Wad Medani
Tel: 83182
Telex: 50009TX80WDSD

Turkey

Eksi, Ismail
Cotton Research Institute
09800 Nazilli
Tel: 3131750
Fax: 3133093

Göven, M. Ali
Plant Protection Research Institute
Diyarbakir
Tel: 2211985
Fax: 2244775

Karaat, Saban
Director
Plant Protection Research Institute
Adana
Tel: 3224639
Fax: 3229520

Mart, Cafer
Plant Protection Research Institute
Adana
Tel: 3219581
Fax: 3229520

Sagir, Abuzer
Plant Protection Research Institute
Diyarbakir
Tel: 2211985
Fax: 2244775

Saydam, Coskun
Director
Plant Protection Research Institute
35040 - Bornova-Izmir
Tel: 3881014
Fax: 3741653

Tamer, Ali
Plant Protection Research Institute
PK 49
06172 - Yenimahalle
Ankara
Tel: 3445993
Fax: 3151531

Tezcan, Füsun
Agricultural Engineer
Plant Protection Research Institute
35040 - Bornova-Izmir
Tel: 3880030
Fax: 3741653

Uzun, Sündüz
Plant Protection Research Institute
35040 - Bornova-Izmir
Tel: 3880030
Fax: 3741653

Zümreoglu, Aydin
Plant Protection Research Institute
35040 - Bornova-Izmir
Tel: 3880030
Fax: 3741653

Yemen
Ba-Angood, Saeed A.
Nasir's College of Agriculture
University of Aden
Khormaksar PO Box 6172
Aden
Tel: 233253
Fax: 232548

Invited speakers
El-Zik, Kamal M.
Professor
Texas A&M University
Department of Soil and Crop
Sciences
College Station
Texas 77843-2474
United States
Tel: 8458263
Fax: 8450456

Braud, Michel
(Coordinator of FAO Interregional
Cooperative Research Network on
Cotton for the Mediterranean and
the Near East)
Le Bours
17380 - Torxé
France
Tel: 597007
Fax: 597792

FAO

Barbosa, Sebastian
Senior Officer, IPM
Food and Agriculture Organization
of the United Nations
Viale delle Terme di Caracalla
00100 Rome
Italy
Tel: 52256269
Fax: 52256347

Doorenbos, J.
FAO Representative
Atatürk Bulvari No. 197
06680 Kavaklidere
Ankara, Turkey
Tel: 4280664, 4270143
Fax: 4274852

M'Boob, Sulayman S.
Senior Regional Crop Protection
Officer
FAO Regional Office for Africa
PO Box 1628
Accra
Ghana
Tel: 666851/4
Fax: 668427

Taher, Mahmoud Mohamed
Regional Plant Protection Officer
for the Near East
FAO Regional Office for the Near
East
11 El Eslah El-Zirai St.
Dokki, Cairo
Egypt
Tel: 3497184
Fax: 3495981

Observer

Iles, Malcolm John
Secretary, Integrated Pest
Management Working Group
Natural Resources Institute
Central Avenue
Chatham Maritime
Kent ME4 4TB
United Kingdom
Tel: 883054
Fax: 883377

FAO TECHNICAL PAPERS

FAO PLANT PRODUCTION AND PROTECTION PAPERS

1	Horticulture: a select bibliography, 1976 (E)
2	Cotton specialists and research institutions in selected countries, 1976 (E)
3	Food legumes: distribution, adaptability and biology of yield, 1977 (E F S)
4	Soybean production in the tropics, 1977 (C E F S)
4 Rev.1	Soybean production in the tropics (first revision), 1982 (E)
5	Les systèmes pastoraux sahéliens, 1977 (F)
6	Pest resistance to pesticides and crop loss assessment – Vol. 1, 1977 (E F S)
6/2	Pest resistance to pesticides and crop loss assessment – Vol. 2, 1979 (E F S)
6/3	Pest resistance to pesticides and crop loss assessment – Vol. 3, 1981 (E F S)
7	Rodent pest biology and control – Bibliography 1970–74, 1977 (E)
8	Tropical pasture seed production, 1979 (E F** S**)
9	Food legume crops: improvement and production, 1977 (E)
10	Pesticide residues in food, 1977 – Report, 1978 (E F S)
10 Rev.	Pesticide residues in food 1977 – Report, 1978 (E)
10 Sup.	Pesticide residues in food 1977 – Evaluations, 1978 (E)
11	Pesticide residues in food 1965-78 – Index and summary, 1978 (E F S)
12	Crop calendars, 1978 (E/F/S)
13	The use of FAO specifications for plant protection products, 1979 (E F S)
14	Guidelines for integrated control of rice insect pests, 1979 (Ar C E F S)
15	Pesticide residues in food 1978 – Report, 1979 (E F S)
15 Sup.	Pesticide residues in food 1978 – Evaluations, 1979 (E)
16	Rodenticides: analyses, specifications, formulations, 1979 (E F S)
17	Agrometeorological crop monitoring and forecasting, 1979 (C E F S)
18	Guidelines for integrated control of maize pests, 1979 (C E)
19	Elements of integrated control of sorghum pests, 1979 (E F S)
20	Pesticide residues in food 1979 – Report, 1980 (E F S)
20 Sup.	Pesticide residues in food 1979 – Evaluations, 1980 (E)
21	Recommended methods for measurement of pest resistance to pesticides, 1980 (E F)
22	China: multiple cropping and related crop production technology, 1980 (E)
23	China: development of olive production, 1980 (E)
24/1	Improvement and production of maize, sorghum and millet – Vol. 1. General principles, 1980 (E F)
24/2	Improvement and production of maize, sorghum and millet – Vol. 2. Breeding, agronomy and seed production, 1980 (E F)
25	*Prosopis tamarugo:* fodder tree for arid zones, 1981 (E F S)
26	Pesticide residues in food 1980 – Report, 1981 (E F S)
26 Sup.	Pesticide residues in food 1980 – Evaluations, 1981 (E)
27	Small-scale cash crop farming in South Asia, 1981 (E)
28	Second expert consultation on environmental criteria for registration of pesticides, 1981 (E F S)
29	Sesame: status and improvement, 1981 (E)
30	Palm tissue culture, 1981 (C E)

Availability: April 1997

Ar	–	Arabic	Multil –	Multilingual
C	–	Chinese	*	Out of print
E	–	English	**	In preparation
F	–	French		
P	–	Portuguese		
S	–	Spanish		

The FAO Technical Papers are available through the authorized FAO Sales Agents or directly from Sales and Marketing Group, FAO, Viale delle Terme di Caracalla, 00100 Rome, Italy.